Ostracoda (Myodocopina) from O‘ahu and French Frigate Shoals, Hawaiian Islands

LOUIS S. KORNICKER is Curator/Research Zoologist and
ELIZABETH HARRISON-NELSON is Museum Specialist
at the National Museum of Natural History,
Smithsonian Institution, Washington, D.C.

S.L. COLES is Research Zoologist,
Department of Natural Sciences,
Bishop Museum, Honolulu, Hawai‘i

Ostracoda (Myodocopina) from O'ahu and French Frigate Shoals, Hawaiian Islands

Louis S. Kornicker, Elizabeth Harrison-Nelson,
and S.L. Coles

Bishop Museum Bulletin in Zoology 8

Bishop Museum Press
Honolulu, 2007

Published by
Bishop Museum Press
1525 Bernice Street
Honolulu, Hawai‘i 96817, USA

ISBN 10: 1-58178-073-7
ISBN 13: 978-1-58178-073-4
ISSN 0893-3146

TABLE OF CONTENTS

ACKNOWLEDGMENTS

The authors thank Ralph DeFelice, Bishop Museum, for sending some of the specimens upon which this paper is based and for correspondence concerning them. All collecting was funded and given logistical support by the U.S. Fish and Wildlife Service. Maps supplied by S.L. Coles were modified by Molly K. Ryan, Smithsonian Institution. Camera lucida drawings by the senior author were inked by Mary Ellen Didion-Carsley of Media Art and Design and Megan Bluhm.

ABSTRACT

Eleven species (10 new) of myodocopid Ostracoda are described from O'ahu, Hawai'i. Five species were collected in Kāne'ohe Bay and 8 in the vicinity of Waikīkī, but only 2 species were common to both locales. The most abundant species in the collections (*Cypridina alpha*, n. sp.) was represented only by adult females, of which most contained either eggs or embryos in their marsupia. Four new species are reported from French Frigate Shoals of which 3 are also present at O'ahu, and one, *Harbansus hox*, is present only at French Frigate Shoals.

INTRODUCTION

Kornicker (1976) described 3 species (*Sarsiella janiceae* Kornicker, *Ancohenia hawaiiensis* Kornicker, and *Microasteropteron youngi* Kornicker) from near Maunalua Bay on the southern coast of O'ahu, and Kornicker (1978: 45) described *Harbansus barnardi* Kornicker from Kāne'ohe Bay, O'ahu; those species are not represented in the present collections from either O'ahu or French Frigate Shoals (Table 1). *Cypridina sharpei* (Müller) was collected in 1902 from surface waters off Honolulu (Sharpe, 1908: 427). That species was referred to "*species indeterminata*" by Kornicker (1991: 28). Two new species are referred to that genus herein. Hartmann (1991: 166, 167) reported 2 species of the suborder Cladocopina Sars (order Halocyprida) from Kaua'i and Maui, Hawai'i. No benthic myodocopid ostracodes have been reported previously from French Frigate Shoals.

The present report on myodocopid Ostracoda from Hawai'i is based on samples collected in November 1999, and January 2000, in Kāne'ohe Bay (Map 1) on the eastern side of O'ahu, and in the vicinity of Waikīkī on the southwestern side of O'ahu (Map 2), and in September 2000, from French Frigate Shoals (Map 3; Appendix 1). Eleven species (10 new) were collected in the samples, 5 in Kāne'ohe Bay and 8 in the vicinity of Waikīkī (Table 1). Only 2 species were common to both localities. Four species were collected from French Frigate Shoals; except for one new species, all are also present at O'ahu (Table 1).

Table 1. Distribution of myodocopid ostracode species in vicinity of O'ahu and French Frigate Shoals.

	O'ahu			French Frigate Shoals
	Kāne'ohe Bay	Waikīkī	Maunalua Bay	
Cypridinidae				
Cypridina alpha, n. sp.	+	+	-	+
Cypridina iota, n. sp.	-	+	-	-
Paravargula trifax Kornicker, 1991	+	-	-	-
Pseudophilomedidae				
Harbansus barnardi Kornicker, 1978	+**	-	-	-
Harbansus hox, n. sp.	-	-	-	+
Cylindroleberididae				
Cylindroleberis kappa, n. sp.	+	-	-	-
Microasteropteron youngi Kornicker, 1976	-	-	+*	-
Parasterope gamma, n. sp.	-	+	-	+
Parasterope iota, n. sp.	-	+	-	-
Parasterope omega, n. sp.	+	+	-	-
Parasterope sigma, n. sp.	+	-	-	-
Parasterope theta, n. sp.	-	+	-	-
Parasterope zeta, n. sp.	-	+	-	+
Synasterope delta, n. sp.	-	+	-	-
Sarsiellidae				
Ancohenia hawaiiensis Kornicker, 1976	-	-	+*	-
Sarsiella janiceae Kornicker, 1976	-	-	+*	-

* Reported in Kornicker (1976: 1), from shallow lagoon between Paiko Peninsula and fringing coral reef in Maunalua Bay.

** Reported in Sampan Pass Kornicker (1978: 45).

METHODS

Similar collection methods, which are described in more detail in Coles *et al.* (2002a, 2002b) and DeFelice *et al.* (2002), were used at the Kāne'ohe Bay, Waikīkī and Maunalua Bay sites on O'ahu and at French Frigate Shoals. A variety of habitats at each site were sampled opportunistically by scraping organisms from hard surfaces and from coral rubble and macroalgae that was returned to the laboratory and held in 70% ethanol before sorting under a dissecting microscope.

A new method of mounting appendages on slides was developed herein: When appendages removed from the body of an ostracode are placed in a drop of glycerine on a slide and then covered with a glass slip, several things may happen: 1, If too little glycerine is used, airspaces form near the edges of the cover slip; 2, if too much glycerine is used, appendages may escape from under the cover slip; 3, if just enough glycerine is used so that the glycerine stops at the edges of the cover slip, some appendages may run out to the edge and be covered by Canada Balsam or another compound used to ring the cover slip. In this study I tried the following technique which eliminated the above problems in almost all cases. Prior to placing the square cover slip over the drop of glycerine containing the appendages, 4 smaller drops of glycerine are placed around the drop containing the appendages (each drop is separated from the central drop by a small space). When the cover slip is applied the drops coalesce and the appendages generally remain some distance from the edges of the slip.

Disposition of specimens. The holotype and most paratypes have been deposited in the Bishop Museum, Hawai'i and have been assigned BPBM numbers. Several paratypes of each species have been deposited in the National Museum of Natural History, Smithsonian Institution, and have been assigned USNM numbers.

Abbreviations. Arrows indicate the anterior. All measurements are in millimeters unless otherwise noted. Roman numerals indicate the endites.

The following abbreviations are used in the legends:

av = anterior view
iv = inside view
lv = lateral view
mv = medial view
nabs = not all bristles shown
ov = outside view
pv = posterior view
vv = ventral view

SYSTEMATICS

Class OSTRACODA Latreille
Subclass MYODOCOPA Sars

The Myodocopa contain the orders Myodocopida Sars and Halocyprida Dana. Only the former is in the present collection.

Order MYODOCOPIDA Sars

The Myodocopida contain the suborder Myodocopina Sars.

Suborder MYODOCOPINA Sars

The Myodocopina contain the superfamilies Cypridinoidea Baird, Sarsielloidea Brady, Norman, and Cylindroleberidoidea Müller. All superfamilies are represented in the present collection.

Superfamily CYPRIDINOIDEA Baird

The Cypridinoidea contain the family Cypridinidae Baird.

Family CYPRIDINIDAE Baird

This family includes 2 subfamilies: Cypridininae Baird and Azygocypridininae Kornicker. Only the former is represented in the collection.

Subfamily CYPRIDININAE Baird

This subfamily include 2 tribes, Gigantocypridinini Hartmann and Cypridinini Baird. Only the latter tribe is represented in the collection.

Tribe CYPRIDININI Baird

The Cypridinini include many genera, of which 2, *Paravargula* and *Cypridina*, are represented in the collection.

Paravargula Cohen & Kornicker

Paravargula Poulsen, 1962: 202. *Nomen nudum*; genus-group name proposed after 1930 without type species designation.

Paravargula Cohen & Kornicker, 1975: 23. **Type species**. *Paravargula ensifer* Poulsen, 1962, original designation.

Composition and distribution. This genus includes 6 species from the Philippines, Singapore, Kei Islands, South Africa, Enewetak (Kornicker, 1991: 5), Taiwan (Tseng, 1977: 178, 191), and O'ahu (herein).

Taxonomic Emendation. Tseng (1977: 178, 191) described two new species of *Paravargula* from Taiwan: *P. taiwantuia* and *P. formosana*. In a key to the species of *Paravargula* presented by Tseng, 1977 (p. 177), *P. formosana* is distinguished from *P. hirsuta* (Müller) by having only 8 claws. Tseng (1977: 191) identified the single specimen on hand as a female. His description of the 7th limb supports the specimen being a juvenile, which would account for the furca having only 8 claws on each lamella, which is a juvenile character (see descriptions of instars of *P. trifax* herein). *Paravargula formosana* is herewith referred to *P. hirsuta*. Fig. 74 in the Tseng 1977 publication is labeled *Paravargula taiwania* [sic] (juvenile), but the specimen number TFM 10276 is the number of the holotype of *P. formosana* given in the description (p. 191). The description of the 7th limb of *P. formosana* (p. 193) accurately describes the 7th limb illustrated in Fig. 74h, but does not accurately describe the 7th limb illustrated in Fig. 72f, which is the illustration referred to in the description of the limb by Tseng (p. 193). Fig. 74 should be labeled *Paravargula formosana*. The legends for Figs. 72 to 74 have errors difficult to resolve.

Paravargula trifax Kornicker
Figs. 1–9

Paravargula sp. Kornicker, 1988: 217.

Paravargula trifax Kornicker, 1991: 5, Figs. 2–4.

Holotype. USNM 158322, adult female on slide and in alcohol.

Type Locality. Enewetak lagoon (night plankton trap, coral reef substrate).

Material. Enewetak: Holotype. Kāne'ohe Bay, O'ahu: Sta 8: BPBM-S 12830, BPBM-S 12831, 2 undissected instars III in alcohol. BPBM-S 12832, adult male, body removed from shell, both in alcohol. Sta 16: BPBM-S 12829, undissected instar V in alcohol. Sta 18: BPBM-S 12823, undissected adult female without eggs in marsupium in alcohol. Sta 20: USNM 1085941, undissected ovigerous female in alcohol; USNM 1085943, adult male in alcohol. Sta 21: USNM 1085942, ovigerous female in alcohol and on slide. Sta 22: USNM 1085943 adult male in alcohol and on slide.

Distribution. Enewetak lagoon, Enewetak Atoll. Kāne'ohe Bay, O'ahu, Hawaiian Islands.

Supplementary description of holotype (Figure 9*c*). Reexamination of the appendage slide of the female holotype (USNM 158322) from Enewetak lagoon described and illustrated by Kornicker (1991: 6; Figures 1–4) revealed the following error in the description of the mandible: 1st article of endopod with 4 (not 3) ventral a-bristles (small bristle with base on lateral side was overlooked). Tip of 7th limb shown in Figure 9*c*.

Description of adult female from O'ahu (Figures 1–6*a,b;* 9*a*). Carapace elongate with convex ventral and dorsal margins, small projecting caudal process and deep incisure (Figure 1*a,c,f;* 2*a*). Surface smooth, with 4 minute bristles just inside outer margin of rostrum (Figure 1*c*), and with abundant minute pores visible at high magnification (x 300) (Figure 1*b*).

Infold: Rostral infold with 15 bristles forming row (some near incisure bifurcate); additional bristles posterior to row (Figure 1*c*); 2 small bristles on infold close to inner end of incisure; about 45 closely spaced divided bristles along anteroventral list and list of anterior half of ventral margin (4 or 5 more posterior bristles more widely spaced); posterior half of ventral margin with about 13 minute bristles. Posterior edge of ridge forming anterior of pocket of caudal process with 19 stout digitate processes and about 8 smaller processes (not more than 1 between each pair of stout processes) (Figure 1*d*); about 12 minute bristles form row posterior to row of stout processes; pocket of caudal process with about 15 minute bristles; about 12 minute bristles just within edge of caudal process.

Selvage: Lamellar prolongation present along anterior and ventral margins, terminating posteriorly near ventral end of caudal process (Figure 1*c,d,f*). Prolongation widest and with narrow striations in vicinity of incisure and divides at inner edge of incisure. A stout process just ventral to incisure parallels striations of prolongation and terminates at distal end of prolongation, and may be part of prolongation (process contains dashes in Figure 1*f*) (this process not described in other members of genus (possibly overlooked) and may act to strengthen prolongation). Inner edge of prolongation located on outer side of valve edge along ventral margin of incisure and along tip of rostrum (Figure 1*f*).

Central Adductor Muscle Attachments (Figure 1*b*): Consisting of about 14 round and elliptical scars.

Hinge: Rounded tooth of left valve fits into U-shaped socket of right valve located just dorsal to end of ridge at anterior end of caudal pocket (Figure 1*d,e*).

Carapace Size (length, height in mm): USNM 1085942: 2.76, 1.27. USNM 1085941: 2.58, 1.35. BPBM-S 12823: 2.52, 1.35.

First Antenna (Figure 2*a*): 1st article bare. 2nd article bare. 3rd article short, with 2 short bristles (1 ventral, 1 dorsal). 4th article with 2 short terminal bristles (1 ventral, 1 dorsal). Sensory bristle of short 5th article with 9 long filaments followed by 3 shorter slender filaments and 1 minute subterminal filament. 6th article with short medial bristle. 7th article: a-bristle similar to bristle of 6th article; b-bristle about 3 times length of a-bristle, with 5 short marginal filaments (some with few spines); c-bristle about twice length of sensory bristle of 5th article, with 8 marginal filaments (with few marginal spines) and 1 shorter subterminal filament. 9th article: d- and e-bristles slightly shorter than sensory bristle of 5th article, bare with oblique tips; f- and g-bristles similar to c-bristle (small spine near base of some filaments). Total bristles 13.

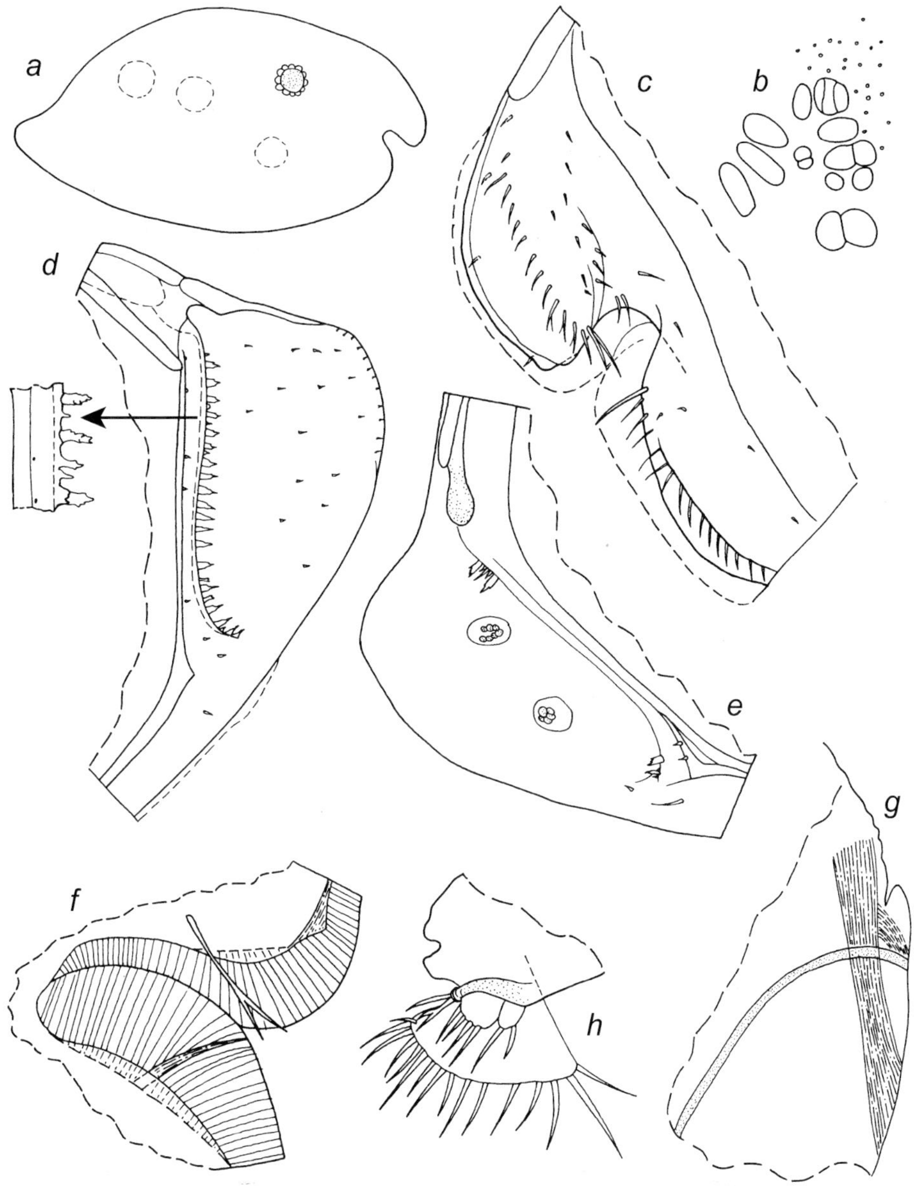

Figure 1. *Paravargula trifax* Kornicker, USNM 1085942, adult female: *a*, complete specimen showing location of central adductor muscle attachments, lateral eye, and 2 eggs in marsupium; length 2.76 mm; *b*, central adductor muscle attachments and pores left valve, ov; *c,d*, anterior and posterior ends right valve, iv; *e*, caudal process left valve (not all teeth shown); 2 disks represent foreign organisms, iv; *f*, incisure left valve, iv; *g*, posterodorsal margin of body from left side showing girdle (stippled) and some muscles (striated); *h*, parts of left 5th and 6th limbs drawn on body, iv (nabs).

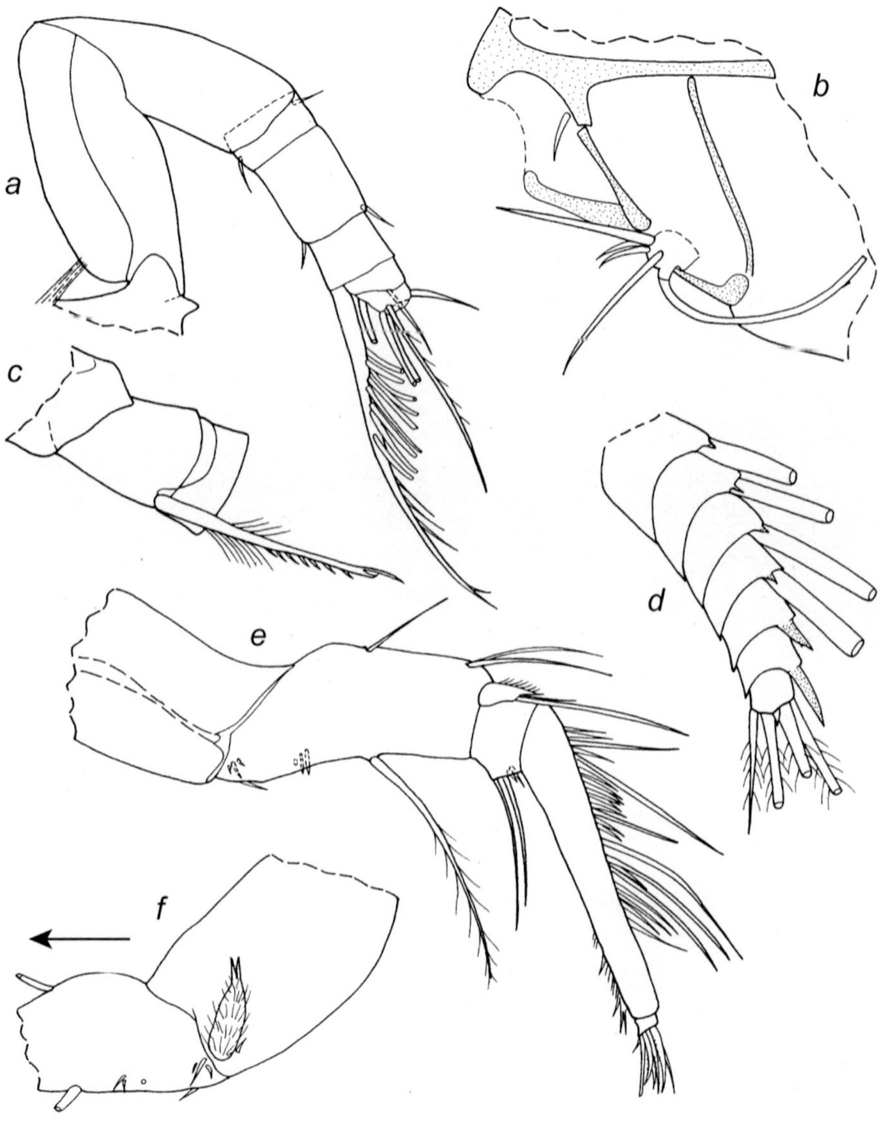

Figure 2. *Paravargula trifax* Kornicker, USNM 1085942, adult female: *a*, right 1st antenna, lv; *b*, part protopod and endopod right 2nd antenna, mv; *c*, ventral bristle of 2nd exopod article right 2nd antenna, mv; *d*, distal part exopod right 2nd antenna, lv (nabs); *e*, right mandible, lv; *f*, proximal part right mandible, mv.

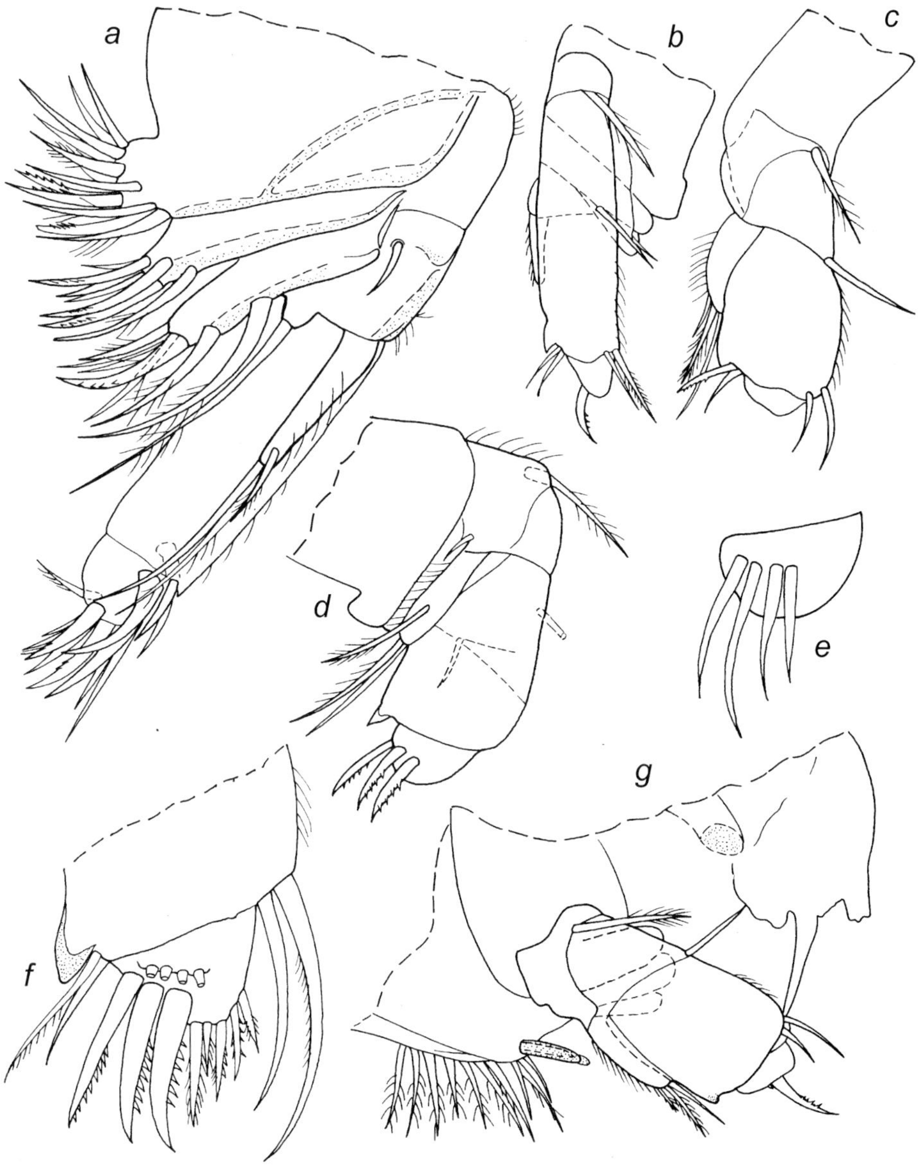

Figure 3. *Paravargula trifax* Kornicker, USNM 1085942, adult female: *a,* right maxilla, (twisted); *b,* right maxilla, (twisted) (nabs); *c,* right maxilla drawn attached to body, lv (nabs); *d,* right maxilla, lv (nabs); *e,* a-bristles right maxilla, lv; *f,* distal part right maxilla, lv (nabs); *g,* view of body from right side showing locations of right 5th limb, right maxilla, mouth (stippled circle), and upper lip (nabs).

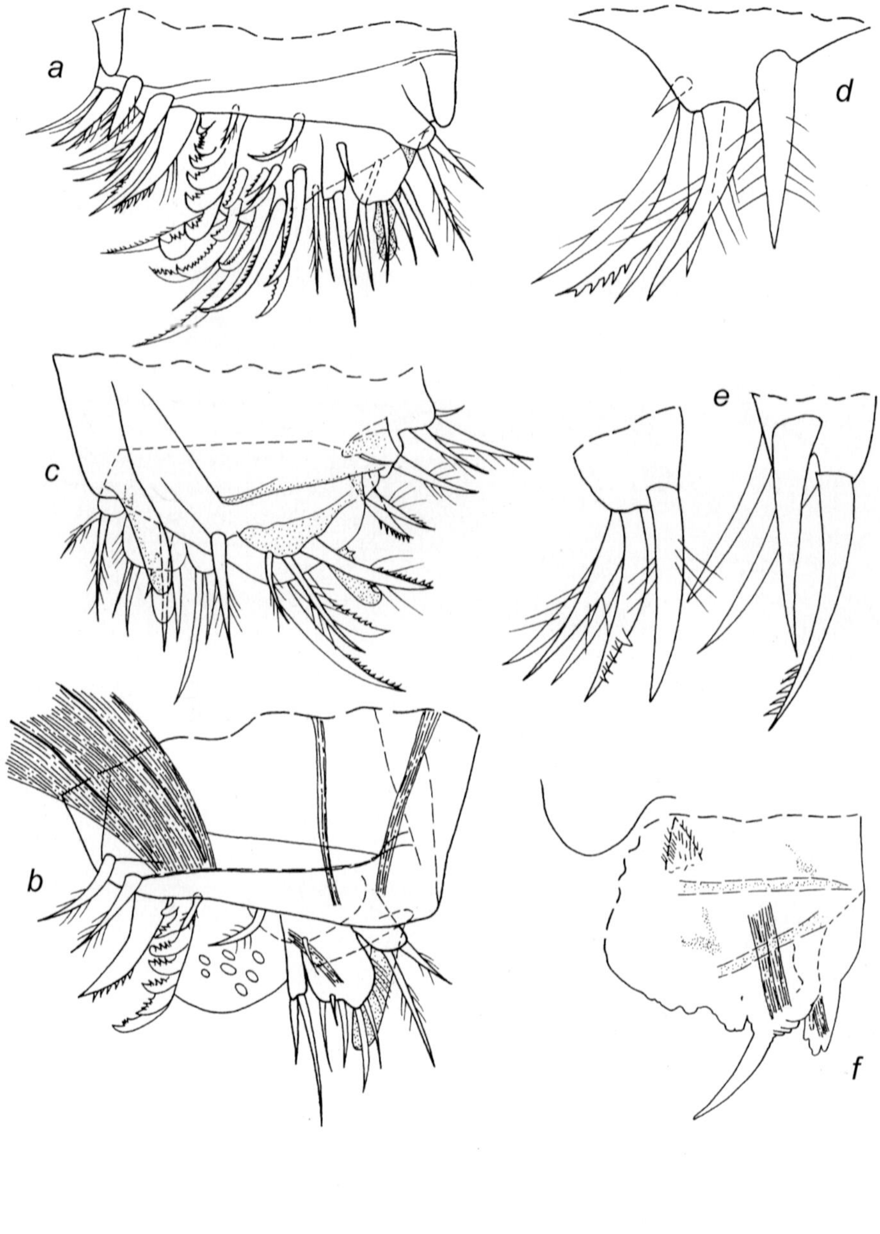

Figure 4. *Paravargula trifax* Kornicker, USNM 1085942, adult female: *a,* distal right 5th limb, pv; *b,* distal right 5th limb showing internal muscles, pv (nabs); *c,* distal right 5th limb, av (nabs); *d,* endite I right 5th limb, pv; **e**, endites II (on left) and III right 5th limb, pv; *f,* right view of body showing curved lower lip (at left), upper lip, and coxa endite of left mandible (dashed triangle with spines).

Second Antenna: Protopodite with small distal medial bristle (Figure 2*b*). Endopod with single article with 4 proximal bristles (2 long, 2 short) and long terminal filament on small protuberance (Figure 2*b*). Exopod: article 1 with minute indistinct spines along concave dorsal margin and along distal ventral margin; 2nd article with indistinct minute spines along ventral margin and bristle reaching 8th article with long proximal hairs followed by 7 ventral spines and short subterminal ringed part (rings not shown) dorsal to unringed part with incurved tip (Figure 2*c*); articles 3 to 8 with long bristle with natatory hairs but no spines; 9th article with 4 bristles: short dorsal bristle with long hairs on left limb, bare on right limb; 3 long bristles with natatory hairs but no spines. No minute spines forming row observed along distal margins of any articles of exopod. Articles 3 to 8 with basal spines; spines on articles 3 to 6 small; spines on articles 7 and 8 large; spine on 8th article about twice length of 9th article (Figure 2*d*); 9th article with small indistinct lateral spine. Total bristles 19.

Mandible (Figure 2*e,f*): Coxa: endite spinous with 2 stout branching terminal spines with small peg between them; small bristle just proximal to base of endite. Basis: dorsal margin with 3 bristles (1 near midlength, 2 subterminal); ventral margin with 4 a-bristles (1 on margin, 3 medial), 2 c-bristles and small unringed peg just proximal to them, and 1 long d-bristles with long proximal and short distal spines. Exopod about three-fourths length dorsal margin of 1st endopod article, with 2 subterminal bristles (distal shorter). Endopod: 1st article with 4 ventral bristles (2 small, 2 long). 2nd article: ventral margin with distal spines and 4 short, unringed, distal bristles; dorsal margin with 9 ringed bristles, 1 short bristle (ringed in proximal part) with short marginal spines (base of bristle just distal and medial to 2nd ringed bristle) and 10 short unringed bare bristles. 3rd article with 3 claws with proximal teeth and 4 ringed bristle (ventral bristle small). Total bristles 48.

Maxilla (Figure 3): Endite I with 11 bristles and claws (Figure 3*a*); endite II with 7 bristles; endite III with 7 distal bristles and 1 small proximal bristle near base of exopod. Coxa with dorsal hairs and spinous dorsal bristle with spines along proximal three-fourths (Figure 3*c,d*). Basis with 1 long dorsal bristle (Figure 3*c*) (broken off and dashed on Figure 3*d*), short medial bristle (dashed on Figure 3*d*), and 1 long, proximal, spinous, ventral bristle near endite III (Figure 3*a*). Exopod with hairs along outer margin, 2 terminal bristles and 1 shorter subterminal bristle (subterminal and outer terminal bristles with long spines). Endopod: 1st article with dorsal hairs and 2 alpha-bristles (outer bristle spinous, other bare), 2 beta-bristles (inner lateral bare, outer medial longer pectinate) and transparent triangular terminal tooth (Figure 3*f*). 2nd article with 4 bare ringed lateral a-bristles (Figure 3*e,f*), 3 b-bristles (anterior ringed, others unringed, clawlike, pectinate), 2 ringed pectinate c-bristles (anterior smaller), and 3 pectinate clawlike d-bristles (Figure 3*d,f*). Approximate total bristles 48.

Fifth Limb (Figures 1*h*, 3*g*, 4*a–e*): Epipod with 48–49 setose bristles. Coxa endite bristles obscured (not all shown in Figure 4*a–c*), endites I to III, each with about 5 bristles (Figure 4*d,e*); distal bristle of endite III broad, pectinate. Long fingerlike protopod tooth with terminal projection. Basis: anterior side with 4 bristles (1 lateral to 3 forming row) (Figure 4*c*); basis endite I (main tooth) with 6 stout pectinate cusps and proximal peg, and short spinous proximal bristle; endite II with 4 pectinate a-bristles, 3 or 4 pectinate b'-bristles, 4 pectinate ringed b"-bristles, short ringed spinous posterior proximal c-bristle, and longer anterior d-bristle with long marginal hairs. Endopod: 1st article with 1 proximal and 2 terminal bristles (shorter bristle without rings); fused articles 2 and 3 with 4 terminal bristles. Exopod with marginal hairs and 2 terminal bristles. Muscles M1 and M2 terminate at exopod, and muscle M3 extends from endopod article 1 and terminates in fused articles 3 and 4 (Figure 4*b*). Approximate total bristles 95.

Sixth Limb (Figures 1*h*, 5*a,c*): Epipod with 4 short bare bristles. Endite I with 5 spinous bristles; endite II with 5 spinous bristles; endite III with 5 spinous bristles; endite IV with 7 spinous bristles. End article with 17 spinous bristles (2 lateral bristles near anterior end short). Total bristles 43.

Seventh Limb (Figures 5*b*, 9*a*): 10 to11 proximal bristles (5 or 6 on each side), each with 4–6 bells; edge of terminal segment (side bearing comb) with 5 bristles, each with 1–7 bells; opposite edge of terminal segment with 6 bristles, each with 1–6 bells. Comb with total of 15 teeth comprising long middle tooth with 1 long slender cusp, 4 short narrow teeth, and 2 short square-tipped teeth on each side; a sclerotized straight bar extends between teeth near bases of teeth. Jaw (Figure 9*a*) consisting of bar with curved bifurcate tip. Total bristles 21 or 22.

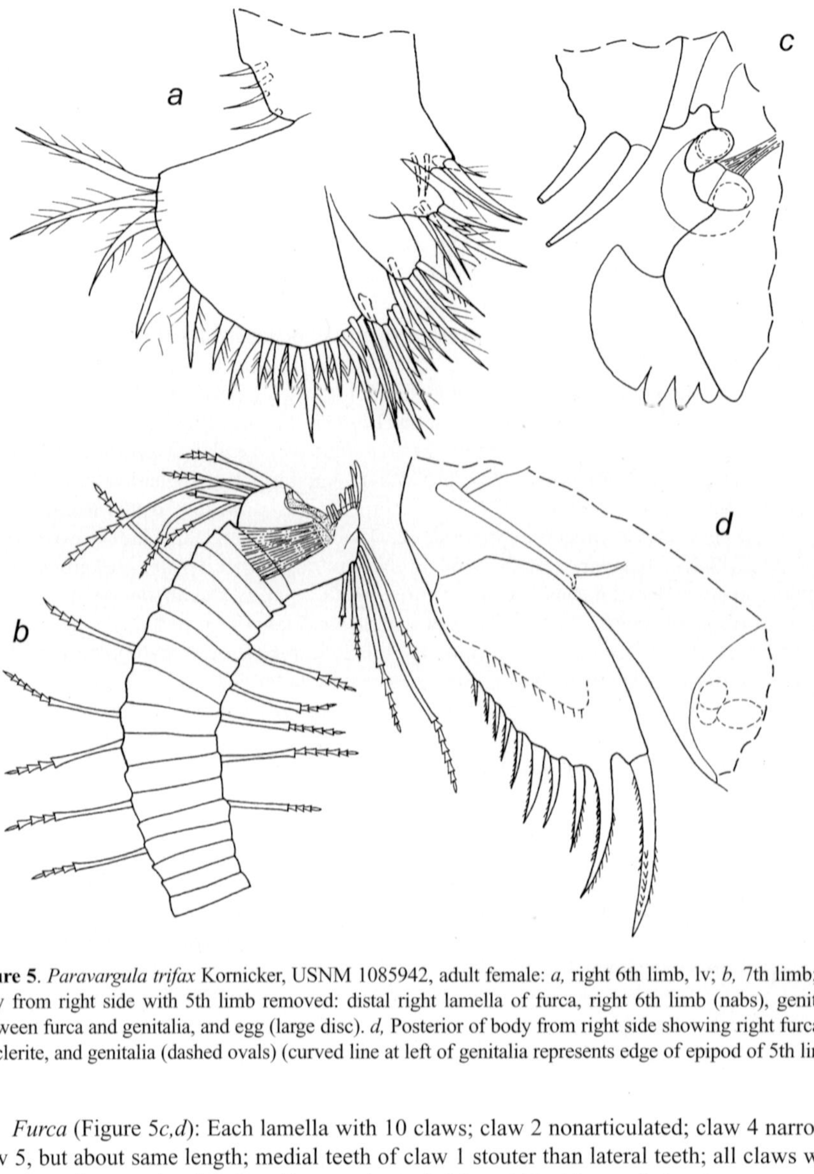

Figure 5. *Paravargula trifax* Kornicker, USNM 1085942, adult female: *a,* right 6th limb, lv; *b,* 7th limb; *c,* part of body from right side with 5th limb removed: distal right lamella of furca, right 6th limb (nabs), genitalia, anus (between furca and genitalia, and egg (large disc). *d,* Posterior of body from right side showing right furcal lamella, Y-Sclerite, and genitalia (dashed ovals) (curved line at left of genitalia represents edge of epipod of 5th limb).

Furca (Figure 5*c,d*): Each lamella with 10 claws; claw 2 nonarticulated; claw 4 narrower than claw 5, but about same length; medial teeth of claw 1 stouter than lateral teeth; all claws with teeth along posterior margin; right lamellar anterior to left by width of base of claw 1.

Belloncі Organ (Figure 6*a*): Short, broad, with triangular tip.

Eyes: Medial eye bare, light brown central part dorsal to dark brown elongate horizontal part with pointed anterior and posterior tips (Figure 6*a*). Lateral eye larger than medial eye, with dense black pigment; about 16 light-amber colored ommatidia visible along edge, some with bisecting suture (Figures 1*a,* 6*a*).

Upper Lip (Figures 3*g,* 4*f,* 6*a,b*): Anterior unpaired part consisting of thin projecting rectangular process with 3 terminal glandular processes (Figure 6*b*). Each side of posterior paired part with long slender tusk with pointed tip and with steplike proximal anterior edge with 2 rows of 4 or 5 glandular processes (total 9), followed by short tusk with 3 terminal glandular processes and a small-

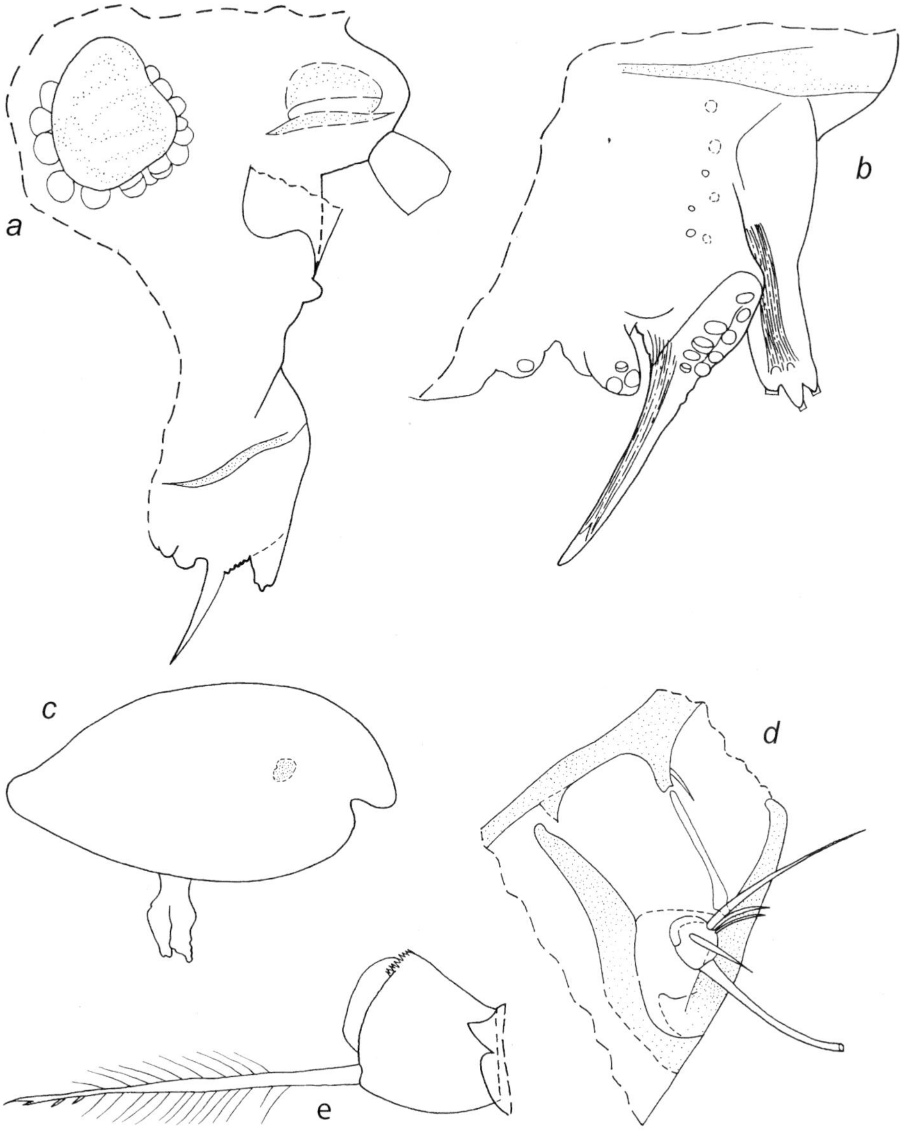

Figure 6. *Paravargula trifax* Kornicker, USNM 1085942, adult female: *a,* anterior of body, from right side with right 2nd antenna removed; *b,* upper lip from right side. USNM 1085943, adult male, length 2.35 mm; *c,* complete specimen showing projecting copulatory organs and right lateral eye, length 2.35 mm; *d,* part protopod and endopod left 2nd antenna, mv; *e,* ventral bristle of 2nd exopod article left 2nd antenna, mv.

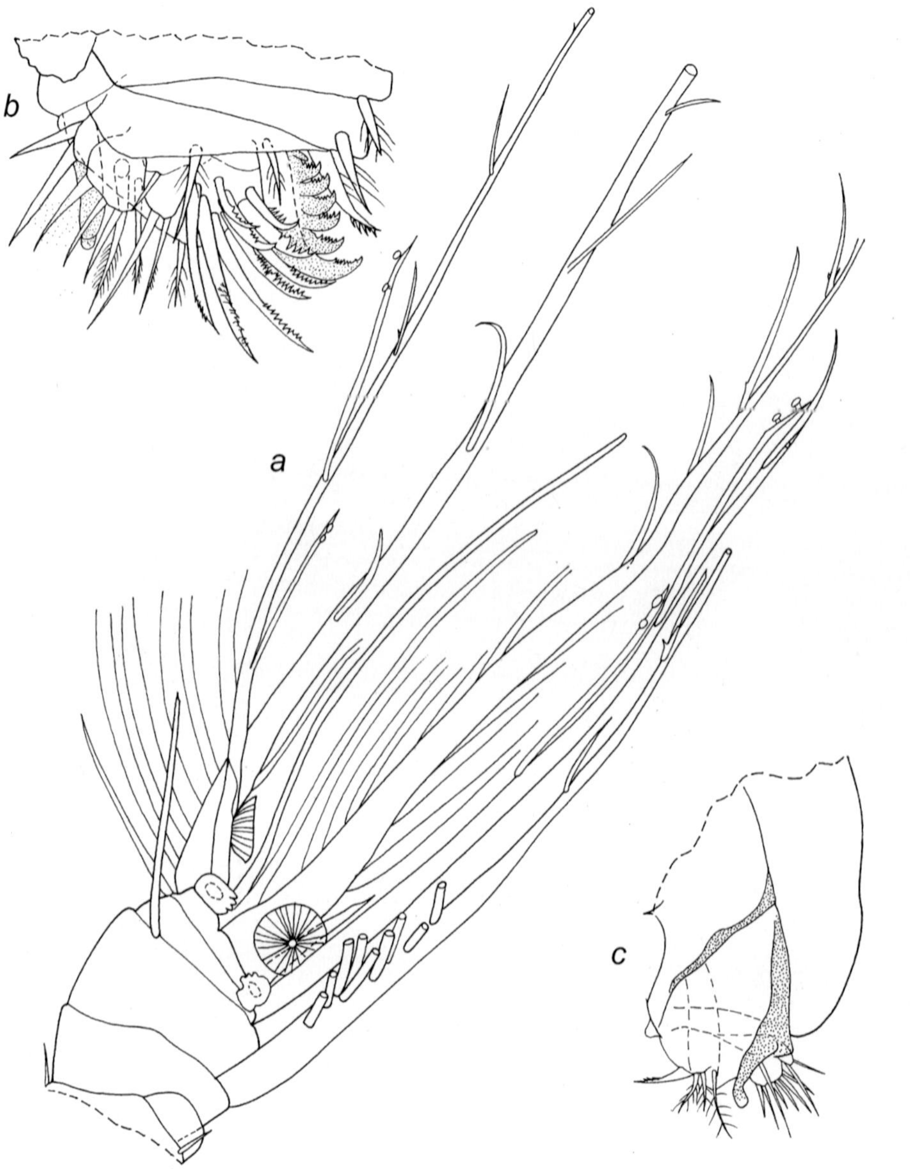

Figure 7. *Paravargula trifax* Kornicker, USNM 1085943, adult male: *a,* distal left 1st antenna, mv; *b,* distal left 5th limb, pv; *c,* part distal left 5th limb, av.

er tusk with 1 glandular terminal process; 3 or 4 minute proximal disks on each side of upper lip may be glandular process, but could not be resolved with certainty.

Anterior of Body (Figure 6*a*): Small rounded process about midway between medial eye and upper lip.

Posterior of Body: Evenly rounded, bare; small triangular process just dorsal to posterior end of girdle (Figure 1*g*).

Y-Sclerite (Figure 5*d*): Typical for subfamily.

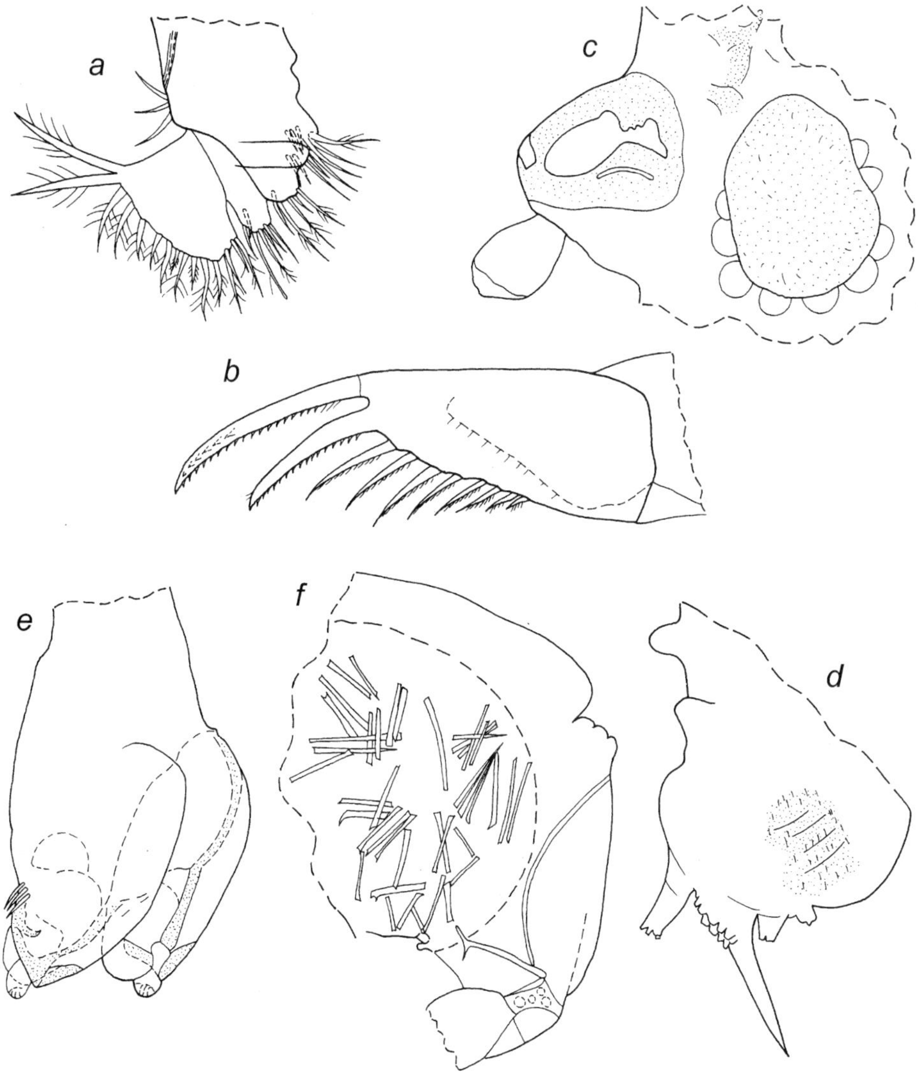

Figure 8. *Paravargula trifax* Kornicker, USNM 1085943, adult male: *a,* right 6th limb, lv; *b,* left furcal lamella, lv; *c,* anterior of body from left side showing left lateral eye, medial eye and Bellonci Organ; *d,* anterior of body showing upper lip; *e,* copulatory organs from left side of body; *f,* posterior of body from left side showing spicules inside gut.

Genitalia (Figure 5*c,d*): Comprising 2 ovoid processes (genitalia and spermatophore) anterior to furca and medial to ventral end of epipod of 5th limb. No genital opening or adjacent bristles observed.

Eggs (Figure 1*a*): USNM 1085942 with 13 round eggs (approximately same diameter) in marsupium; diameter of 1 egg 0.241 mm. USNM 1085941 with about 37 round eggs (approximately same diameter) in marsupium; diameter of 1 egg 0.259 mm.

Gut Content: USNM 1085942 with unrecognized particulate matter.

Foreign organisms: USNM 1085942 with 2 transparent oval organisms inside pocket of caudal process of left valve (Figure 1*e*); each transparent oval contains several amber colored cells.

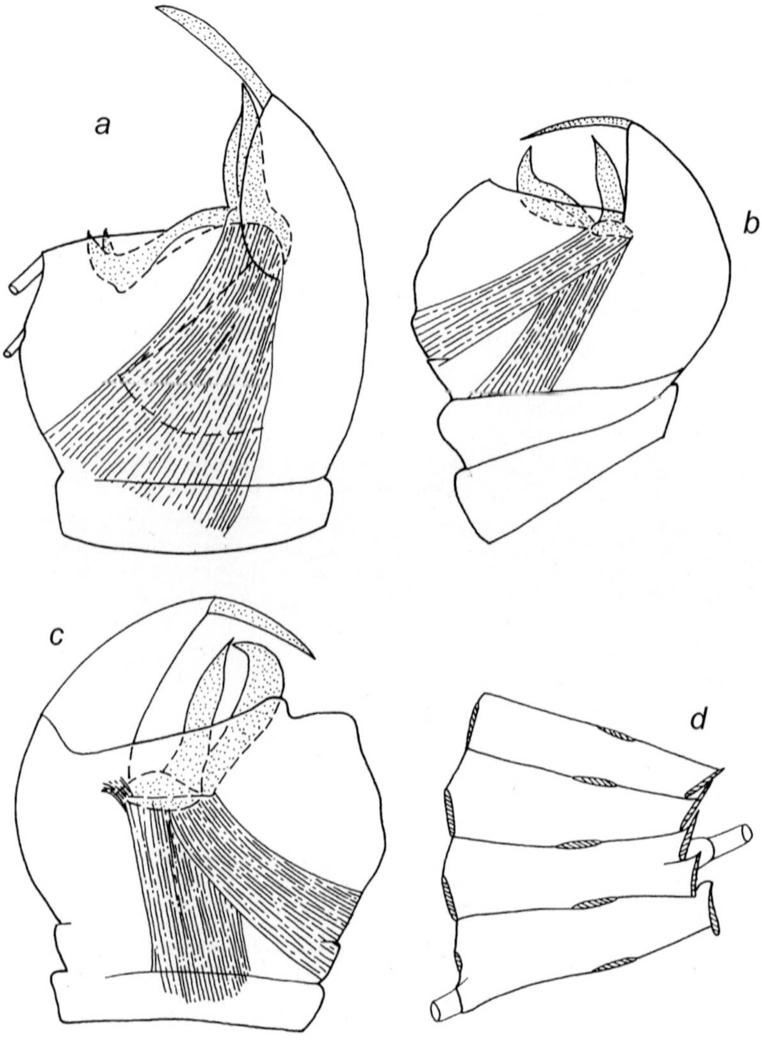

Figure 9. *Paravargula trifax* Kornicker, tip of 7th limbs: *a*, USNM 1085942, adult female; *b*, USNM 1085943, adult male; *c*, holotype, USNM 158322, adult female from Enewetak Atoll. *d*, USNM 1085943, adult male, distal rings of 7th limb

Description of adult male (Figures 6*c–e,* 7, 8, 9*b,d*). Carapace shorter than adult female and with a more oblique posterodorsal margin (Figure 6*c*).

Carapace Size (length, height in mm): USNM 1085943: 2.35, 1.23. BPBM-S 12823: length only, 2.41. BPBM-S 12832: length only 2.17.

First Antenna (Figure 7*a*): Articles 1 to 4 similar to those of adult female. Sensory bristle of 5th article stout, with 9 long proximal filaments, 3 short distal filaments with proximal spine, and 1 minute subterminal filament. 6th article with short slender medial bristle. 7th article: a-bristle similar to bristle of 6th article, b-bristle slightly longer than sensory bristle, with proximal pointed process with 1 large sucker, then 2 long slender filaments with 2 small subterminal suckers, followed by 2 short filaments each with a basal spine; bristle distal to last short filament with minute marginal spines. c-bristle almost 3 x as long as b-bristle, with proximal process with larger sucker (with greater diameter than proximal sucker of b-bristle), then 2 longer slender filaments each with 2 small subterminal suckers, followed by 6 long filaments (each with basal spine) and 1 minute filament near tip. 8th article: d- and e-bristles bare, shorter than b-bristle; f- and g-bristles same length as c-bristle, with abundant long proximal hairs (not all shown in illustration) and few distal minute marginal spines, and with 9 slender filaments (each with basal spine) and 1 minute filament near tip.

Second Antenna: Protopod similar to that of adult female (Figure 6*d*). Endopod of left limb differs from that of adult female in having 1 long and 3 short bristles, and in terminal filament not being on projection (Figure 6*d*); endopod of right limb same as that of adult female. Exopod: 2nd article with minute spines along inner part of distal margin (Figure 6*e*); bristle of article 2 with only 2 to 4 distal ventral spines proximal to subterminal spine (Figure 6*e*); short dorsal bristle of 9th article with long marginal hairs on both left and right limbs. Limb otherwise similar to that of adult female.

Mandible: Coxa, basis, and 1st endopod article same as that of adult female. 2nd article of endopod: ventral margin from midlength to base of paired subterminal bristles with small distal spines, and 4 short, unringed, distal bristles; dorsal margin with 9 ringed bristle, 1 short bristle ringed in proximal half and with short marginal spines (base of bristle just distal to 2nd long ringed bristle) and 15 short unringed bare bristles. 3rd article with 3 claws and 4 ringed bristle (ventral bristle small). Total bristles 53.

Maxilla: Endite I with 12 bristles; endite II with 7 bristles; endite III with 7 bristles and 1 small bristle near base of exopod. Remainder of limb similar to that of adult female. Total bristles 49.

Fifth Limb (Figure 7*b,c*): Epipod with 50 plumose bristles. Remaining limb same as that of adult female. Approximate total bristles 96.

Sixth Limb (Figure 8*a*): 3 short bristles in place of epipod. Endite I with 6 bristles (3 terminal, 3 proximal); endite II with 5 bristles (3 terminal, 2 proximal); endite III with 5 bristles (4 terminal, 1 proximal); endite IV with 7 bristles (6 terminal, 1 proximal). End article with 16 or 17 bristles (2 or 3 lateral bristles near anteroventral corner short). Total bristles 42 or 43.

Seventh Limb (Figure 9*b,d*): 11 proximal bristles (6 on jaw side, 5 on comb side), each with 4–6 bells; edge of terminal segment (side bearing comb) with 6 bristles, each with 1–7 bells; opposite edge of terminal segment with 6 bristles, each with 1–7 bells. Comb with total of 11 teeth comprising long middle cusp with 1 slender cusp, 3 short narrow teeth, and 2 short square-tipped teeth on each side. Jaw consisting of bar with curved tip. Total bristles 23.

Furca (Figure 8*b*): Each lamella with 10 claws similar to those of adult female, except claw 5 not as broad at base.

Bellonci Organ (Figure 8*c*): Similar to that of adult female.

Eyes: Lateral eye slightly larger than that of adult female, with black pigment and many ommatidia (Figure 8*c*). Medial eye similar to that of adult female, but with more black pigment (Figure 8*c*).

Upper Lip (Figure 8*d*): Similar to that of adult female.

Anterior of Body (Figure 8*d*): Small rounded process on anterior margin between upper lip and medial eye.

Posterior of Body (Figure 8*f*): Evenly rounded bare. Several small bulges dorsal to posterior end of girdle.

Y-Sclerite (Figure 8*f*): Similar to that of female, both with short terminal branches.

Copulatory organ (Figure 8*e*): Elongate, paired, with 3 bristles.

Gut Content: USNM 1085943 and BPBM-S 12832 with numerous unidentified amber colored rodlike sclerites with hooked tips (Figure 8*f*).

Description of instar III (sex undetermined). Carapace similar in lateral outline to that of adult female.

Carapace Size (length, height in mm): BPBM-S 12830, 1.12, 0.56. BPBM-S 12831, 1.18, 0.50.

Sixth Limb: End segment with many bristles.

Seventh Limb: Long without bristles.

Furca: Each lamella with 7 claws. Claw 2 fused to lamella. Claw 5 stouter and longer than claw 4.

Foreign Organisms: BPBM-S 12831 with numerous transparent oval discs within pocket of caudal processes of each valve. Each oval contains amber colored cells.

Description of instar V (sex undetermined). Carapace similar in lateral outline to that of adult female.

Carapace Size (length, height in mm): BPBM-S 12829, 1.86, 0.96.

Seventh Limb: Elongate with bristles, some tapered.

Furca: Each lamella with 9 claws. Claw 2 fused to lamella; claw 5 longer and stouter than claw 4.

Mechanics of seventh limb: The tip of the 7th limb of *P. trifax* bears a short stout muscle similar to that described in the 7th limb of *P. hirsuta* Müller, 1906 by Poulsen (1965: 209): "A large, but short, transverse muscle passes from the dorsal margin of the limb to the corner between the proximal parts of the jaw and the comb." The tips of 2 females and one male, which is smaller than that of the females, are shown in Figure 9*a–c*. The muscle (or muscles) of the female limb in Figure 9*a* is interpreted to be in the relaxed state. The central tooth of the comb points distally, the elongate pointed tooth at the base of the comb is flat against the base, and the jaw opposite the comb is approximately at a right angle to the comb. The muscle of the male limb in Figure 9*b* is interpreted to be partly contracted. The central tooth of the comb no longer points distally, the elongate pointed tooth has been pulled away from the base of the comb, and the jaw has moved slightly towards the elongate pointed tooth. The muscle of the female limb in Figure 9*c* is interpreted to be completely contracted. The location of the central tooth of the comb is only slightly changed, but the elongate pointed tooth has separated farther from the base of the comb, and the tip of the jaw is in contact with the elongate pointed tooth. The ability of the elongate pointed tooth and jaw to go from an open to closed position suggests that it may have a clasping function.

The limb proximal to the last article consists of numerous narrow rings. Except for the distal approximately 13 rings, the sutures separating the rings have a uniform width. The distal 13 sutures separating the rings have a short, wide, sclerotized section midway between the bases of the bristles on each side of the limb (only top section shown in Figure 9*d*). The 7th limb is quite flexible. At the concave edge of a bend the rings overlap, whereas, at the convex side, the rings appear to stretch (Figure 9*d*).

Comparisons. The specimens from O'ahu differ from those from Enewetak in having a dense black pigment between ommatidia of the lateral eye. The description of *P. trifax* by Kornicker (1991: 5) was based on a single specimen from Enewetak lagoon, an adult female. Other species of Myodocopa collected from Enewetak are also without pigment in the lateral eyes. Because of this I interpret the absence of pigment in the lateral eyes of *P. trifax* from Enewetak to be the result of preservation. Parts of the body of the Enewetak specimen had an underlayer of black pigment absent on the specimens from O'ahu. Also, the female of *P. trifax* from Enewetak is smaller (length 2.44 mm) than those from O'ahu (lengths in mm: 2.76, 2.58, 2.52). Those differences, at least that of size, are attributed to intraspecific variability.

Genus *Cypridina* Milne-Edwards

Cypridina Milne-Edwards, 1840: 409. Poulsen, 1962: 255. Kornicker, 1987: 173; 1991: 27; 1992: 11. **Type species**: *Cypridina renaudii* Milne-Edwards, 1840, by monotypy.
Cypridina (Pyrocypris) Müller, 1890: 230.
Eupathistoma Brady, 1898: 437.
Pyrocypris Müller, 1890: 213; 1906: 16; 1912: 16.
Cypridina (Cypridina): Skogsberg, 1920: 313.

Composition. Including 2 new species described herein, 27 species are recognized as species of *Cypridina* sensu Poulsen (1962: 255) but only 18 are sufficiently known to recognize at the species level (Kornicker, 1991: 27, 1992: 11).

Distribution. Indian and Pacific Oceans between latitudes of about 35°N and 30°S, planktonic and demersal (Kornicker, 1991: 28).

Taxonomic Emendation. Tseng (1977: 99, 102) described 2 new species from Taiwan Strait: *C. taiwania* and *C. chinai*. Tseng (1977: 59) presented a key to species of *Cypridina*, but did not include the 2 new species in the key. Both species are described in detail (p. 99, 102), but comparisons with other species are minimal. In the description of the shell of *C. taiwania*, Tseng (p. 99) stated, "The form of shell is quite different from the others, in the mid-anterior dorsal margin with a notch distinctly." It seems possible that the "notch" is an artifact developed after death of the animal. However, the senior author examined 16 specimens of *C. alpha* new species, immersed in glycerin on a dished slide, and none had such a dorsal notch, so possibly the notch on *C. taiwania* is a specific difference, but I am doubtful, and herewith refer *C. taiwania* to species inquirenda.

In the description of *C. chinai* Tseng (1977: 111) stated, "The shell is similar to those of *C. inermis* but slightly smaller than *C. inermis*." This difference is insufficient to define the species. Being unable to identify with certainty the specimens described by Tseng, I herewith refer *C. chinai* to species inquirenda.

Cypridina alpha, new species

Figs. 10–17

Holotype. BPBM-S 12852, ovigerous female on slide and in alcohol.

Type Locality. Sta 7, 18 Jan 2001, Canoes, Waikīkī, O'ahu.

Paratypes. Kāne'ohe Bay: Sta 8: USNM 1085944, 1 ovigerous female. Sta 10: BPBM-S 12833, 1 ovigerous female; BPBM-S 12834, 1 ovigerous female; BPBM-S 12835, 1 ovigerous female; USNM 1085945, 1 ovigerous female; USNM 1085946, 1 ovigerous female. Sta 17: BPBM-S 12824, 1 ovigerous female. Sta 21: BPBM-S 12822, 1 ovigerous female. Sta 24: BPBM-S 12842, 1 ovigerous female.

Waikīkī: Sta 7: BPBM-S 12853, 8 ovigerous females. Sta 8: USNM 1085947, 19 ovigerous females; BPBM-S 12854, 1 ovigerous female. Sta 9: USNM 1085948, 2 adult females (1 ovigerous); BPBM-S 12858, adult female without eggs. Sta 10: BPBM-S 12859, 51 adult females (44 ovigerous, 2 with embryos, 5 without eggs). Sta 11: USNM 1085949, 1 female with embryos; USNM 1085950, 1 ovigerous female; BPBM-S 12860, 16 ovigerous females; USNM 1085949b, USNM 1085949c, USNM 1085951, 3 embryos from USNM 1085949.

Non-types. French Frigate Shoals: Sta TC-7: BPBM-S 12843, 3 females. Sta TC-8: BPBM-S 12844, 26 females. Sta TC-9: BPBM-S 12845, 1 female. Sta TC-10: BPBM-S 12846, 3 females. Sta TC-12: BPBM-S 12847, 1 female. Sta TC-13: USNM 1085952, 2 females. Sta RJ-15: BPBM-S 12850, 1 female. Sta RJ-17: BPBM-S 12851, 1 female.

Distribution. Waikīkī: Sta 7–11. Kāne'ohe Bay: Sta 8, 10, 17, 21, 24. French Frigate Shoals: TC-7–10, -12, -13; RJ-15, -17.

Description of adult female (Figures 10–15*a–c, e–h*). Carapace elongate with convex ventral and dorsal margins and well-developed projecting caudal process with rounded posterior margin (Figures 10*a–f,* 11*a,* 14*a,e*). Anterodorsal corner of rostrum evenly rounded, inferior tip of rostrum pointed, anterior margin of rostrum slightly concave, and with deep incisure. Outer surface of valve smooth, with 1 minute bristle at midheight of outer margin of rostrum, 4 small bristles forming row ventral to inner end of incisure and setback from anterior edge of valve, and small bristles along ventral edge of valve. Outer surface of anterior edge of valve just ventral to incisure with 4 or 5 indistinct broad triangular inward pointed spines (Figure 10*b,e,f*). Anteroventral edge of each valve with 7

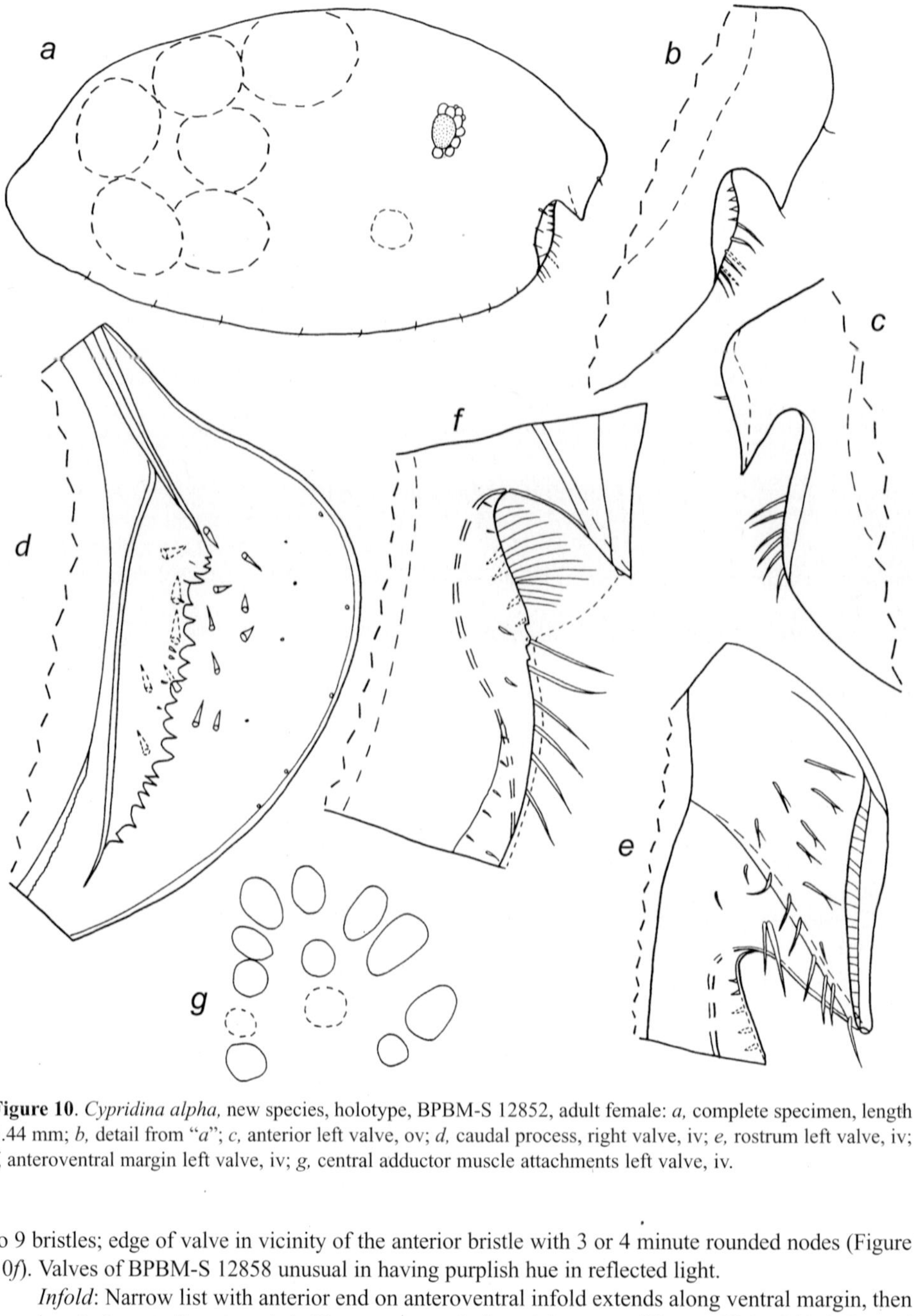

Figure 10. *Cypridina alpha,* new species, holotype, BPBM-S 12852, adult female: *a,* complete specimen, length 1.44 mm; *b,* detail from "*a*"; *c,* anterior left valve, ov; *d,* caudal process, right valve, iv; *e,* rostrum left valve, iv; *f,* anteroventral margin left valve, iv; *g,* central adductor muscle attachments left valve, iv.

to 9 bristles; edge of valve in vicinity of the anterior bristle with 3 or 4 minute rounded nodes (Figure 10*f*). Valves of BPBM-S 12858 unusual in having purplish hue in reflected light.

Infold: Narrow list with anterior end on anteroventral infold extends along ventral margin, then broadens to form to form broad shelf at anterior end of caudal process (Figures 10*d,e,f,* 11*a*). Rostral infold with 15 bristles (many divided) plus pair of bristles at inner end of incisure and 2 small bristles proximal and posterior to pair of bristles (Figure 10*e*); about 35 short bristles along anteroventral and ventral list. Posterior edge of ridge forming anterior of pocket of caudal process of right

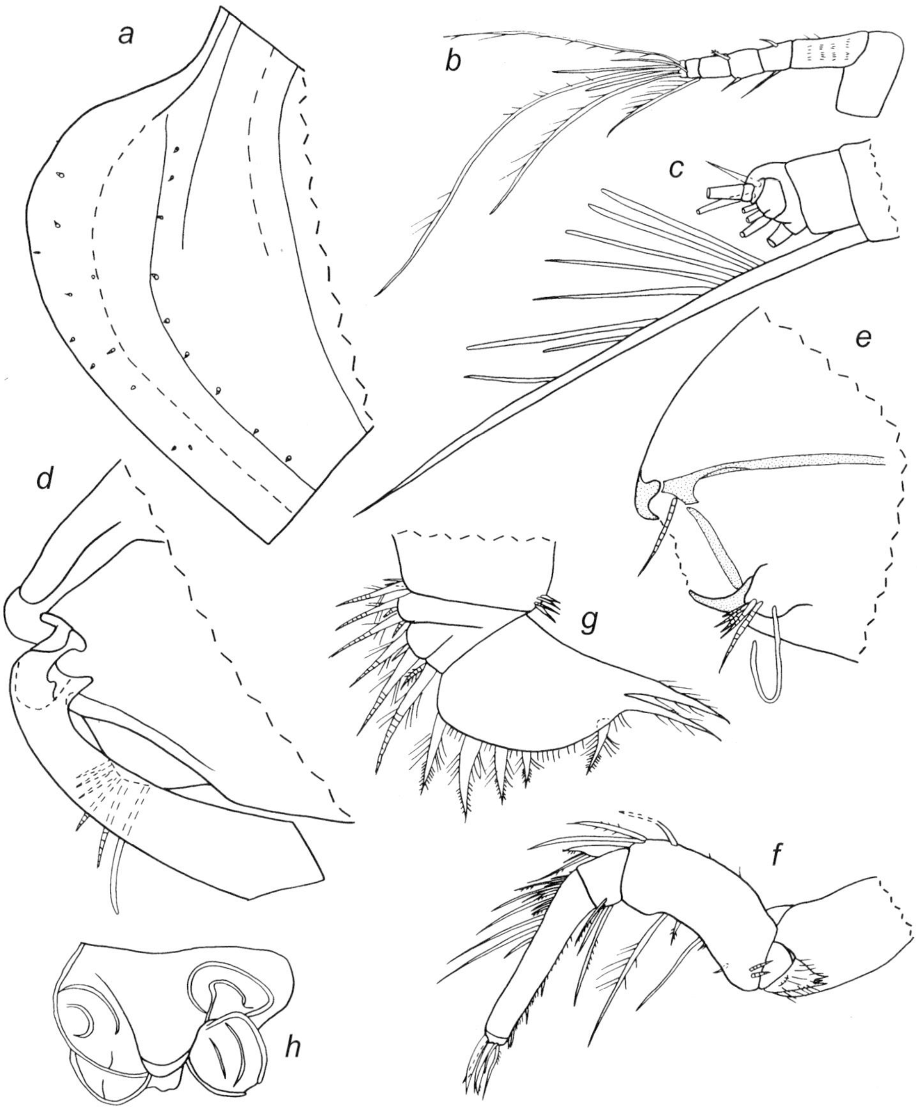

Figure 11. *Cypridina alpha,* new species, holotype, BPBM-S 12852, adult female: *a,* caudal process left valve, iv; *b,* right 1st antenna, mv; *c,* tip left 1st antenna, lv; *d,* part left 2nd antenna, lv; *e,* part right 2nd antenna, mv; *f,* right mandible, mv; *g,* left 6th limb, lv; *h,* genitalia removed from body.

valve with 19 or 20 stout spines (Figure 10*d*); posterior edge of ridge of left valve smooth (Figure 11*a*); 9 minute bristles form row anterior to posterior edge of left ridge. Pocket of caudal process of only right valve with 14 or 15 stout bristles, some beneath broad anterior ridge (Figure 10*d*). Both valves with minute bristles present just within edge of caudal process; few minute bristles or pores present in pocket.

Selvage (Figure 10*f*): Lamellar prolongation with smooth outer edge present along anterior and ventral margins, terminating posteriorly near ventral end of caudal process. Prolongation widest and with narrow striations in vicinity of incisure and divides at inner edge of incisure.

Central Adductor Muscle Attachments (Figure 10*g*): Consisting of about 12 round and elliptical scars.

Carapace Size (length, height in mm): Kāne'ohe: Sta 21: BPBM-S 12822, 1.27, 0.73. Sta 17: BPBM-S 12824, 1.29, 0.74. Sta 24: BPBM-S 12842, 1.28, 0.72. Sta 10: BPBM-S 12833, 1.23, 0.74; BPBM-S 12834. 1.24, 0.72; BPBM-S 12835, 1.28, 0.75; USNM 1085945, 1.25, 0.74; USNM 1085946, 1.20, 0.66.

Waikīkī: Sta 7: BPBM-S 12852, 1.44, 0.77. Sta 8: BPBM-S 12854, 1.41, 0.77. Sta 9: USNM 1085948, 2 specimens, 1.39, 0.77; 1.50, 0.82; BPBM-S 12858, 1.49, 0.84. Sta 10: BPBM-S 12859, 1.44, 0.81. Sta 11: USNM 1085949, 1.47, 0.85; USNM 1085950, 1.50, 0.80.

French Frigate Shoals: Sta RJ-17: BPBM-S 12851, 1.40, 0.79.

First Antenna (Figures 11*b,c;* 14*a,f*): 1st article bare. 2nd article with ventral, dorsal, and medial spines. 3rd article short, with 2 short bare bristles (1 ventral, 1 dorsal). 4th article with 2 short terminal bristles (1 bare ventral, 1 spinous dorsal). Sensory bristle of long 5th article with 6 long filaments followed by 2 shorter more slender filaments (proximal of these close to base of distal long filament). 6th article with short bare medial bristle. 7th article: a-bristle bare, slightly longer than bristle of 6th article; bare b-bristle shorter than d- and e-bristles, with proximal rings; c-bristle about twice length of sensory bristle of 5th article, with 7 or 8 marginal filaments (some with few marginal spines). 8th article: d- and e-bristles about same length as sensory bristle of 5th article, bare with oblique tips (e-bristle shorter than d-bristle); f-bristles long with about 8 marginal filaments; g-bristle longer than f-bristle, with about 11 marginal filaments.

Second antenna (Figures 11*d,e;* 14*a–d,* 15*e*): Protopodite with small medial bristle (Figure 11*d*). Endopod with single article with 6 bristles (1 long (filamentlike), 1 medium (ringed), 4 short (ringed)) (Figures 11*c,d,* 14*b–d*). Exopod (Figure 14*a–d*): article 1 bare or with few spines along concave dorsal margin; 2nd article with bristle reaching 7th article, with 9 ventral spines (last spine longer); articles 3 to 8 with long bristle with natatory hairs but no spines; 9th article with 3 bristles: short dorsal bristle with few short hairs; 2 long bristles with natatory hairs but no spines. No minute spines forming row observed along distal margins of any articles of exopod. Articles 3 to 8 with slender basal spines (not shown); spines longer on distal articles; spine on 8th article almost as long as 9th article; 9th article with pointed lateral spine slightly longer than 9th article (not shown).

Mandible: (Figures 11*f,* 14*a*): Coxa: endite spinous with 2 stout branching terminal spines with smaller triangular process between them; small bristle at base of endite. Basis: dorsal margin with few proximal spines and 3 bristles (1 distal to midlength, 2 subterminal); proximal ventral corner with 2 small medial a-bristles; ventral margin with 1 small b-bristle, 2 c-bristles (1 long, 1 short), and 2 spinous d-bristles. Exopod about same length as dorsal margin of 1st endopod article, with 2 subterminal bristles (distal shorter), and hirsute tip. Endopod: 1st article with 4 ventral bristles (1 long, 2 short, 1 minute). 2nd article: ventral margin with spines and 4 spinelike bristles (2 single, 2 terminal paired); dorsal margin with few proximal spines, 5 short spinous bristles, 3 medium length bristles, and 4 long bristles. 3rd article with 3 claws of equal length (ventral and middle claw with proximal ventral teeth) and 4 ringed bristles (3 near ventral margin (long ventral bristle S-shaped in proximal part and with proximal ventral spines, others short medial with short spines) and 1 lateral near dorsal margin).

Maxilla (Figure 12*a–c*): Endite I with 7 bristles and claws (Figure 12*b*); endite II with 5 bristles (Figure 12*b*); endite III with 5 distal bristles (Figure 12*b*). Precoxa with fringe of dorsal hairs (Figure 12*c*). Coxa with dorsal hairs and spinous dorsal bristle (Figure 12*a*) (bristle missing, probably broken off right limb of holotype (Figure 12*c*)). Basis with 1 or 2 short bristles. Exopod with hairs along outer margin, 2 terminal bristles and 1 shorter subterminal bristle (subterminal and outer of terminal bristles with long spines). Endopod: 1st article with dorsal hairs and 1 alpha-bristle, 2 beta-bristles and 2-pronged cutting tooth. 2nd article with 2 or 3 ringed lateral a-bristles, 3 b-bristles, 2 c-bristles, and 3 pectinate clawlike d-bristles.

Fifth Limb (Figures 12*d–f,* 13*a*): Epipod with 35 setose bristles. Coxa endite I with 4 bristles (Figure 12*e*); endite II with 5 bristles and 1 triangular process; endite III with 6 bristles and 1 triangular process. Basis: anterior side with 4 bristles (1 lateral to 3 forming row) (Figure 13*a*); basis

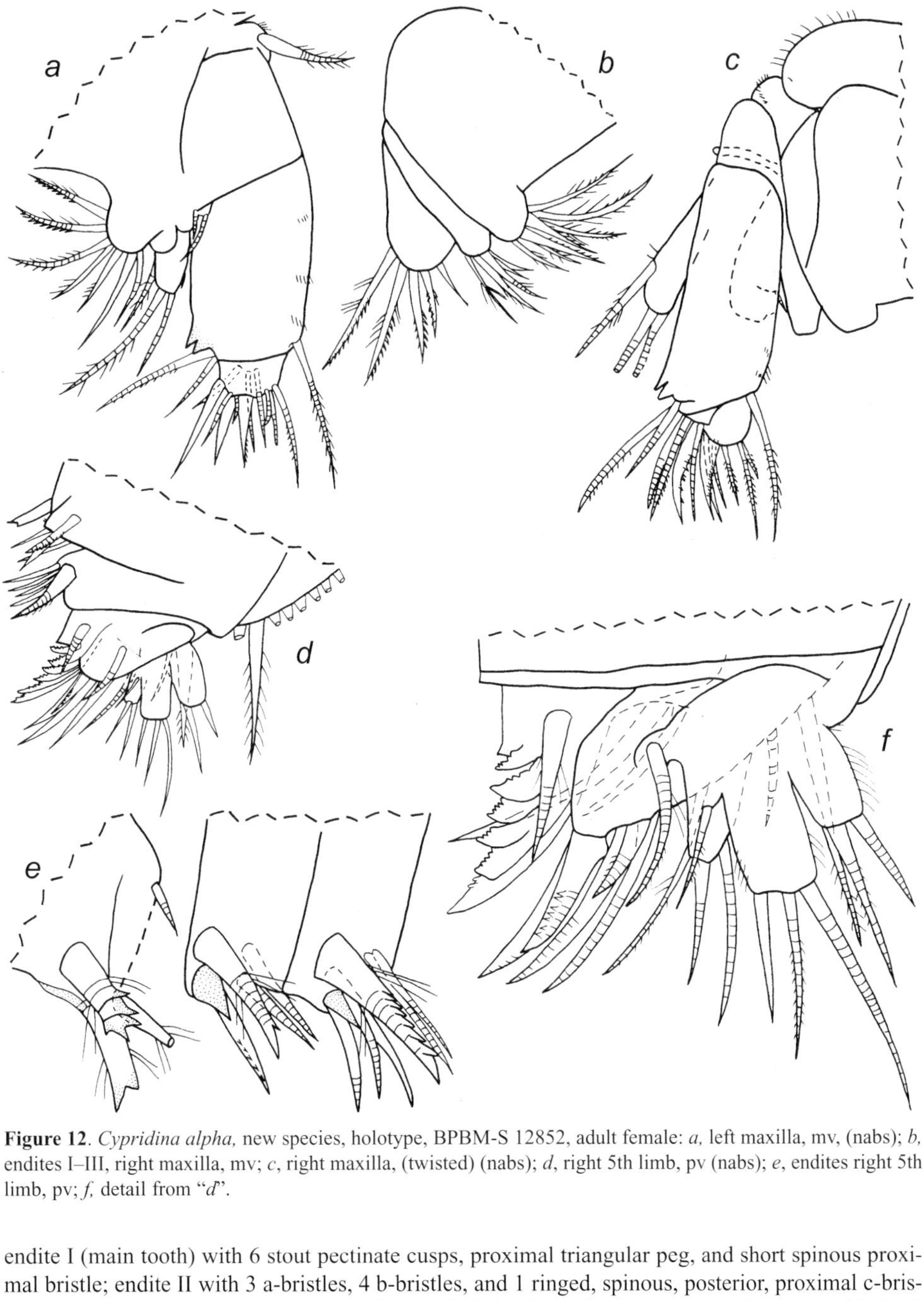

Figure 12. *Cypridina alpha,* new species, holotype, BPBM-S 12852, adult female: *a,* left maxilla, mv, (nabs); *b,* endites I–III, right maxilla, mv; *c,* right maxilla, (twisted) (nabs); *d,* right 5th limb, pv (nabs); *e,* endites right 5th limb, pv; *f,* detail from "*d*".

endite I (main tooth) with 6 stout pectinate cusps, proximal triangular peg, and short spinous proximal bristle; endite II with 3 a-bristles, 4 b-bristles, and 1 ringed, spinous, posterior, proximal c-bristle; anterior side of endite II with long hirsute d-bristle (dashed) adjacent to smaller bristle (dashed) (Figure 12*f*) (small bristle absent on left limb of holotype (Figure 13*a*)). Endopod: 1st article with 1 proximal bristle and 2 terminal bristles (shorter terminal bristle without rings) (Figures 12*f,* 13*a*); fused articles 2 and 3 with 3 terminal bristles. Exopod with marginal hairs and 2 terminal bristles.

Sixth Limb (Figures 11*g,* 15*e*): Epipod with 4 short bare bristles. Endite I with 1 long and 2

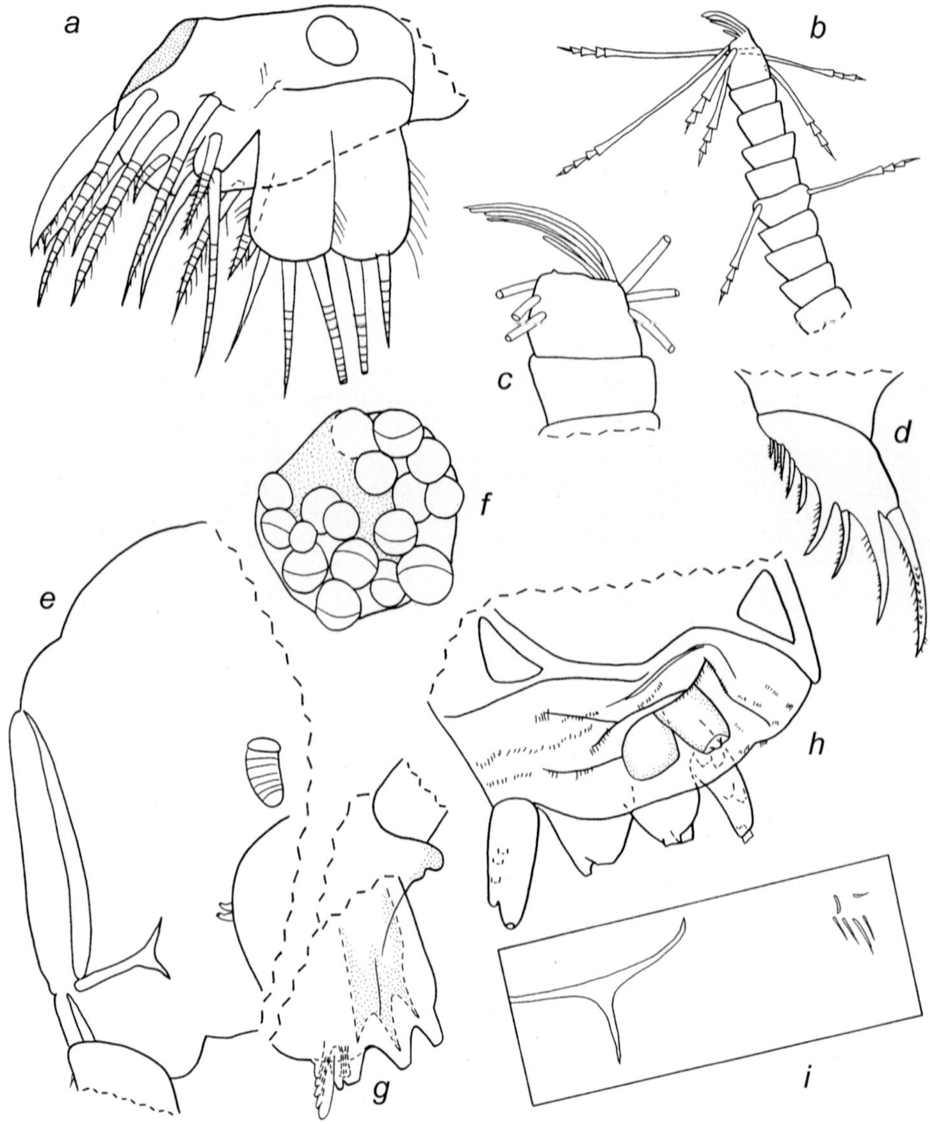

Figure 13. *Cypridina alpha,* new species, holotype, BPBM-S 12852, adult female: *a,* distal left 5th limb, av; *b,* 7th limb; *c,* tip 7th limb; *d,* right lamella of furca, lv; *e,* posterior of body showing parts of 5th limb, 7th limb, furca, Y-Sclerite, and girdle; *f,* lateral eye; *g,* anterior of body from right side showing upper lip, anterior process, and proximal right 1st antenna; *h,* posterior view of upper lip crushed under cover slip; *i,* posterior of body from right side showing anterior part Y-Sclerite and genital bristles.

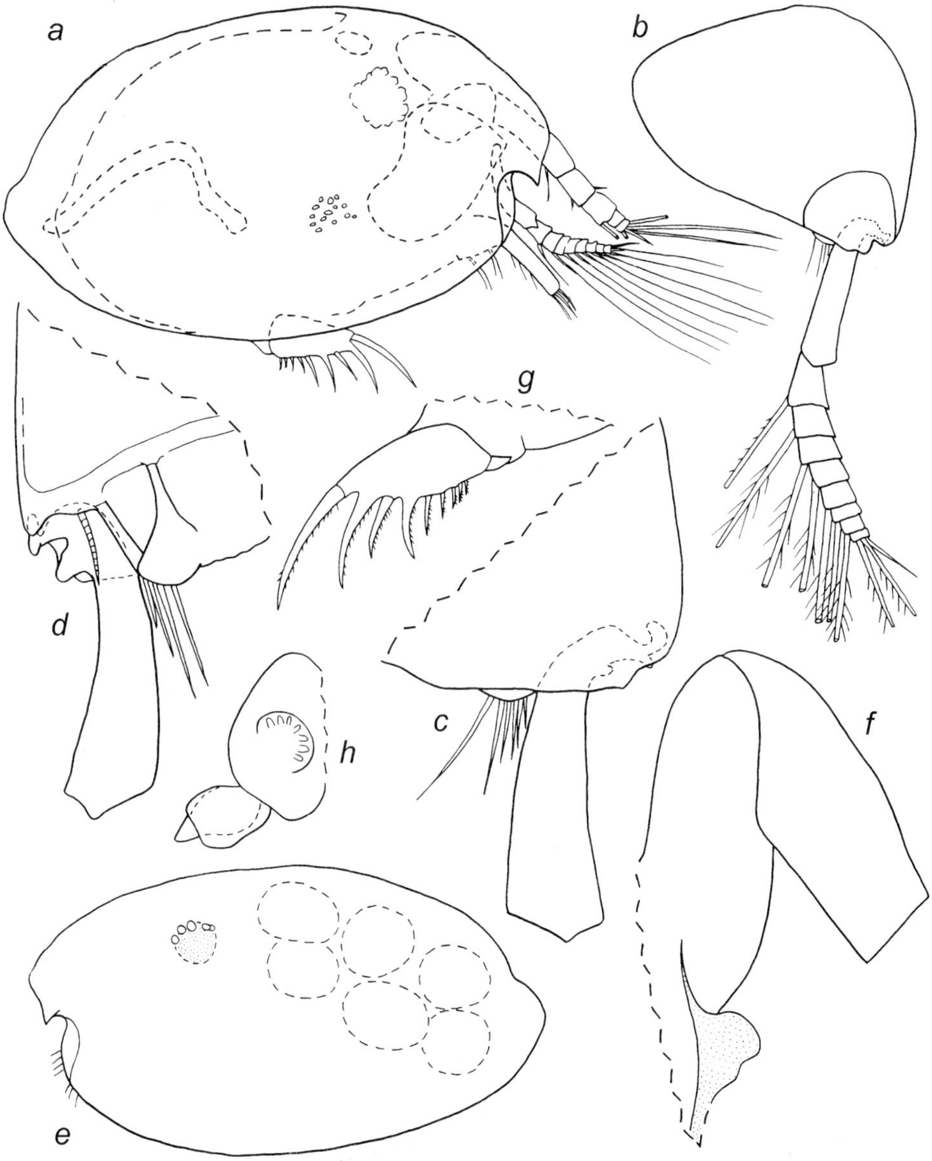

Figure 14. *Cypridina alpha,* new species, paratype, BPBM-S 12824, adult female: *a,* complete specimen, length of carapace 1.29 mm; *b,* right 2nd antenna, lv; *c,* detail from "*b*"; *d,* part right 2nd antenna, mv. Paratype, BPBM-S 12822, adult female: *e,* complete specimen showing eggs and lateral eye, length 1.27 mm; *f,* part left 1st antenna, mv, and anterior process (stippled); *g,* left lamella of furca, lv; *h,* medial eye and Bellonci Organ.

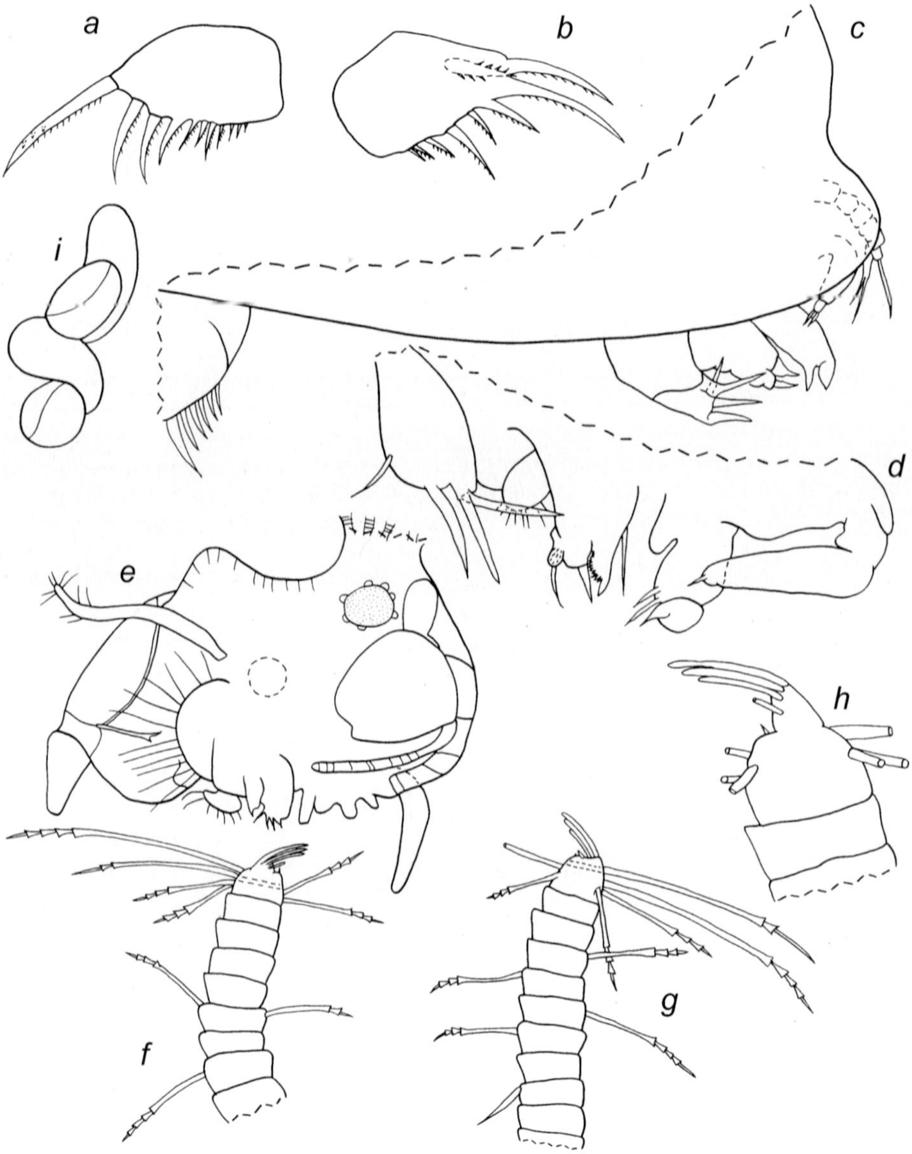

Figure 15. *Cypridina alpha,* new species, paratype BPBM-S 12858, adult female: *a,* right lamella of furca, mv; *b,* left lamella of furca, mv. Paratype, USNM 1085949, adult female containing embryos: *c,* posteroventral part of carapace showing part of projecting furca and embryo (USNM 1085954); *d,* detail of embryo in "*c*"; *e,* body of female removed from carapace (nabs); *f,* 7th limb; *g,* 7th limb opposite that shown in "*f*"; *h,* detail from "g"; *i,* both genitalia removed from body.

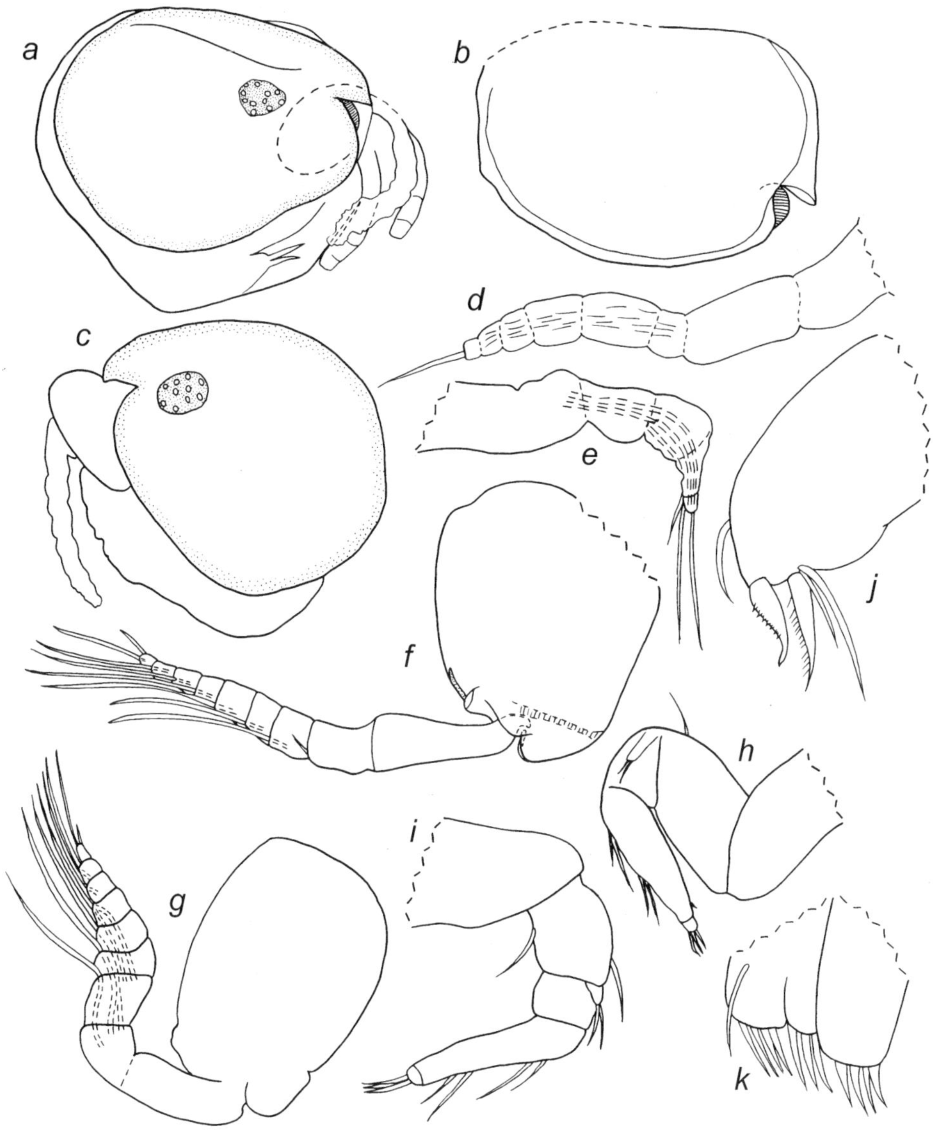

Figure 16. *Cypridina alpha,* new species, embryo (instar I?) USNM 1085949c, from inside carapace of paratype USNM 1085949, adult female: *a,* complete specimen from right side, carapace length 0.40 mm; *b,* left valve, iv, length 0.43 mm; *c,* complete specimen from left side showing only left 2nd antenna; *d, e,* 1st antennae viewed under cover slip (unextruded bristles dashed); *f,* left 2nd antenna, mv; *g,* right 2nd antenna, lv; *h, i,* left and right mandibles, lv; *j,* distal endopod of maxilla; *k,* endites of maxilla.

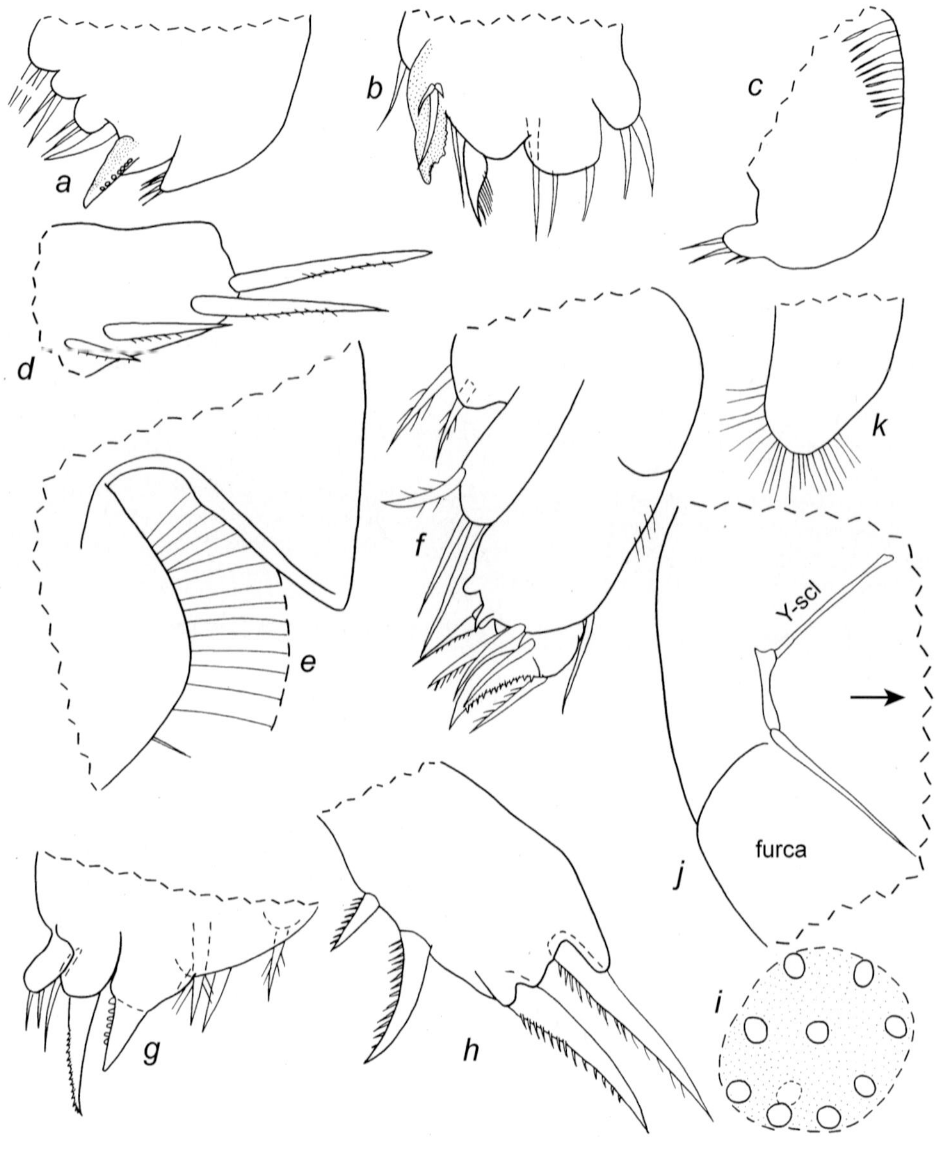

Figure 17. *Cypridina alpha,* new species, embryo (instar I?) USNM 1085949c, from inside carapace of paratype USNM 1085949, adult female: *a, b,* distal parts of 5th limbs; *c,* posterior of body from left side showing left lamella of furca and series of lines in posterodorsal part of body; *d,* right lamella of furca. Embryo (instar I?), USNM 1085949b from inside carapace of USNM 1085949: *e*, rostrum, incisure, and lamellar prolongation of selvage of right valve, ov (complete carapace length 0.41 mm); *f,* right maxilla, lv (nabs); *g,* distal parts right 5th limb, lv; *h*, right lamella of furca, lv; *i*, right lateral eye; *j,* posterior of body from left side showing Y-Sclerite and girdle; *k*, right 6th limb, lv.

short bristles; endite II with 2 long and 2 short bristles; endite III with 2 long bristles and 1 or no short bristle; endite IV with 1 long and 1 short bristle. End article with 6 short and 2 long bristles, stiff spines along lateral edge, and abundant hairs on medial side.

Seventh Limb (Figures 13*b,c,* 14a, 15*e–h*): 2 short proximal bristles (1 on each side), each with 3 bells; edge of terminal segment with 6 bristles, 3 on each side (2 short bristles on each side with 3 bells, 1 long bristle on each side with 4 bells) (Figure 13*b,c*). Comb with total of 7 teeth (middle tooth long, 2 on each side slight shorter; 1 on each end short; under oil immersion (x 100), long bristles appear to have thick walls except at tips. Minute triangular process opposite comb (Figures 13*c,* 15*h*). Variability: 3 specimens (BPBM-S 12852, AA, PPP) with 2 proximal bristles on each limb (Figure 13*b*); 1 specimen (USNM 1085949) with 3 proximal bristles on one limb (2 on comb side with 3 bells, 1 on other side with 2 bells) (Figure 15*f*), and 5 on other limb (2 on comb side with 3 or 4 bells, 3 on other side (1 proximal with no bells, 2 with 3 bells) (Figure 15*g*).

Furca (Figures 13*d,* 14*a,g,* 15*a,b*): Each lamella with 9 claws with teeth along posterior edges; claws 2 and 4 nonarticulated; claw 3 narrower than claw 4, but about same length; claw 1 with stout medial teeth and with few distal anterior spines; right lamellar anterior to left by width of base of claw 1. Furca of BPBM-S 12858 aberrant in having left lamella anterior to right by width of base of claw 1, one of the claws only partly extruded, and left and right lamellae with a different arrangement of fused claws (Figure 15*a,b*).

Bellonci Organ (Figure 14*h*): Short, broad, with triangular tip.

Eyes: Medial eye bare, without distinct coloration (Figure 14*h*). Lateral eye larger than medial eye, with dense black pigment and about 17 light amber-colored ommatidia, some with bisecting suture (Figure 13*f*).

Upper Lip (Figures 13*g,h,* 15*e*): With 2 unpaired anterior processes and 2 paired posterior processes.

Anterior of Body (Figures 13*g,* 14*f*): Small rounded process ventral to base of 1st antenna.

Posterior of Body (Figure 13*e*): Evenly rounded, bare.

Y-Sclerite (Figure 13*e,i*): Typical for subfamily.

Genitalia (Figures 11*h,* 15*i*): Comprising 2 ovoid processes (each with attached spermatophore) anterior to furca and medial to ventral end of epipod of 5th limb.

*Brush*like *Bristles* (Figure 13*i*): 5 small bristles anterior to anterior end of Y-Sclerite.

Eggs (Figures 10*a,* 14*e*): BPBM-S 12852 with 10 eggs; lengths of 2 eggs (mm): 0.276, 0.268. BPBM-S 12822 with 13 eggs; lengths of 2 eggs (mm): 0.232, 0.233. BPBM-S 12854 with 7 oval eggs; maximum lengths of 3 eggs (mm): 0.316, 0.324, 0.325. BPBM-S 12859 with 10 eggs. USNM 1085949 with 12 embryos in marsupium (some measurements given below).

Gut Content: BPBM-S 12822 with crustacean fragments.

Description of embryo (Figures 15*c,d,* 16, 17). Embryos from brood chamber of USNM 1085949. Soft body projects out of carapace along anterior, posterior, and ventral, and margins (Figure 16*c*). Bristles on many appendages only partly emergent. All appendages, body, and valves filled with abundant small round cells. No instars encased. Details of appendages difficult to resolve.

Carapace: Carapace oval in lateral view when viewed under cover slip (Figure 16*b*). Incisure well developed but projecting caudal process present in adult absent (Figure 16*a–c*). Bristles along anteroventral corner of adult absent. Abundant small round organic cells present on inner sides of valves.

Infold (Figure 16*b*): Weakly developed but present along free margins. Bristles absent. Spines not present on infold of caudal process of right valve.

Selvage (Figures 16*b,* 17*e*): Striated broad lamellar prolongation present along anterior margin ventral to incisure (similar to that of adult). A prolongation also visible projecting anteriorly from tip of rostrum.

Central Adductor Muscle Attachments: Obscured by numerous cells present on inner surface of valves.

Carapace Size (length, height in mm): Waikīkī, Sta 11: USNM 1085949c: measured in drop of

Table 2. Distribution of barren and ovigerous *Cypridina alpha* females, and of lateral eyes in eggs and embryos in marsupia of ovigerous females in samples from Kāne'ohe Bay and Waikīkī, O'ahu, and French Frigate Shoals. "Eyes present" and "Eyes absent" refer to eyes in eggs.

	Barren females	Ovigerous females		
Stations (date)		Eyes absent Eggs	Eyes present Eggs	Embryos
Kāne'ohe Bay				
8 (12 Nov 1999)	0	1	0	0
10 (12 Nov 1999)	0	5	0	0
17 (8 Nov 1999)	0	1	0	0
21 (5 Nov 1999)	0	1	0	0
24 (25 Jan 2000)	0	1	0	0
Total specimens	0	9	0	0
Percent specimens	0	100	0	0
Waikīkī				
7 (18 Jan 2001)	0	4	5	0
8 (18 Jan 2001)	0	18	2	0
9 (22 Jan 2001)	2	1	0	0
10 (22 Jan 2001)	5	40	4	2
11 (22 Jan 2001)	0	12	5	1
Total specimens	7	75	16	3
Percent specimens	6.9	74.3	15.8	3
Kāne'ohe Bay and Waikīkī Combined				
Total specimens	7	82	17	4
Percent specimens	6.4	74.5	15.5	3.6
French Frigate Shoals				
FFS-TC-7 (11Sep 2000)	0	1	0	0
FFS-TC-8 (12 Sep 2000)	1	22	3	0
FFS-TC-9 (12 Sep 2000)	0	1	0	0
FFS-TC-10 (12 Sep 2000)	2	1	0	0
FFS-TC-12 (13 Sep 2000)	1	1	0	0
FFS-TC-13 (14 Sep 2000)	1	0	0	0
FFS-R-15 (23 Sep 2000)	0	1	0	0
FFS-R-17 (23 Sep 2000)	1	0	0	0
Total specimens	6	27	3	0
Percent specimens	16.7	75	8.3	0

glycerine 0.40, 0.30; left valve measured compressed under cover slip 0.42, 0.31. USNM 1085949, right valve measured compressed under cover slip 0.41, 0.28. USNM 1085951 measured in drop of glycerine 0.41, 0.32. Outlines of instars in lateral view including parts of body projecting from valves, irregular, triangular, or rounded (total length USNM 1085949c 0.49 mm, total height 0.38 mm).

First Antenna (Figure 16*d,e*): Sutures indicated by indentations in ventral and dorsal margins, but sutures mostly not visible. Bristles only partly emergent, bare.

Second Antenna (Figure 16*c, f,g*): Protopod without distal medial bristle. Endopod bare. Exopod with 9 weakly divided articles. Article 1 with indistinct short bare bristle. Articles 3 to 8 with partly emerged bare bristles. Article 9 with 2 indistinct bare bristles. Basal spines absent. Lateral spine of article 9 absent.

Mandible (Figure 16*h,i*): Bristles indistinct. Coxa endite absent. Exopod with 2 terminal bristles. Endopod article 3 with 3 bristles (2 claw-shaped).

Maxilla: With 3 endites (Figure 16*k*). Endopod article 3 with pectinate terminal claws (Figures 16*j*, 17*f*). Exopod well developed with 3 bristles (Figure 17*f*).

Fifth Limb (Figure 17*a,b,g*): Epipod with bristles present. Basis: endite I (main tooth) with 1 stout tooth with marginal crenulations. Endites II and III well developed with marginal bristles (exact number not resolved). Exopod well developed with 2 terminal bristles.

Sixth Limb (Figure 17*k*): Small with marginal hairs.

Seventh Limb: Absent.

Furca (Figures 15*c,d*, 17*c,d,h*): Each lamella with 4 claws with teeth forming row along posterior edges.

Bellonci Organ: Not observed, possibly obscured.

Eyes: Lateral eye with dark pigment and 10 or 11 separated ommatidia (Figures 16*a,c*, 17*i*). All 12 instars from USNM 1085949 have lateral eyes with dark pigment. Medial eye not observed, possibly obscured.

Lips: Present, lobate.

Posterior of Body (Figure 15*e*): Evenly rounded.

Y-Sclerite (Figure 17*j*): Without anterior branch.

Genitalia: Absent.

Remarks. None of the embryos were enclosed by a membrane, nor were membranes present inside the adult carapace after the instars were removed. Poulsen (1962: 36–41, 69–71, 96, 97) described several embryos carried within the shell of 3 species of *Gigantocypris*.

Comparisons. The adult male of the species is unknown. The adult female is characterized by having claws 2 and 4 of the furca nonarticulated, in having posterior spines along the anterior ridge of the caudal process of the right valve and no spines on the left valve, and in having 7 to 9 bristles along the anteroventral edge of each valve and 3 or 4 low nodes in the vicinity of the anterior bristle. The female *C. spinula* Kornicker and *C. segrex* Kornicker have furcae with claws 2 and 4 nonarticulated.

The new species is very close to *C. mellentini* Kornicker & Harrison-Nelson described from Tutuila, American Samoa, and could be a subspecies. The observed differences in the 2 species is in small morphological differences in the carapaces. Both species have spines along the anterior ridge of the caudal process of the right valve and none on the left, but *C. alpha* has 19 or 20 stout spines, whereas, *C. mellentini* has only 10 to 13 smaller spines. Also, the pocket of the caudal process of the right valve of *C. alpha* has 14 or 15 stout bristles compared to 2 at most on *C. mellentini.* The edge of each valve of *C. alpha* in the vicinity of the anterior bristle in the row of bristles along the anteroventral margin has 3 or 4 minute rounded nodes absent on *C. mellentini.* The anteroventral margin of the valves of *C. alpha* have 4 or 5 inward pointing spines absent on *C. mellentini.* The lengths of valves of the adult female *C. melletini* range from 1.14 to1.25 mm, average 1.20 mm (9 specimens) (Kornicker & Harrison-Nelson, 2006: 246), whereas the lengths of valves of the adult female *C. alpha* range from 1.20 to 1.50 mm, average 1.36 mm (16 specimens). However, the lengths of specimens of *C. alpha* from Kāne'ohe, Oahu, average 1.26 mm (8 specimens) compared to 1.46 mm (8 specimens) for specimens from Waikīkī, Oahu, indicating that the size of that species varies considerably in different localities. Those species differ from *C. alpha* in having spines along the posterior edge of the anterior ridge of the caudal process of the left valve, and in not having a row of 7–9 bristles along the anteroventral edge of each valve.

Composition of populations (Table 2). All specimens of *C. alpha* in the samples were adult females. The percentages of females with eggs or embryos were 100% for Kāne'ohe Bay, 93.1% for Waikīkī, and 83.3% for French Frigate Shoals. The reason for the unusual concentration of adult and ovigerous females is unknown. Samples were collected in November 1999, January 2000, and September 2000. Three specimens from Waikīkī (2.9%) contained embryos. Possibly the embryos could be interpreted to be at the instar I stage. Black lateral eyes were present in eggs of 17.3% of the ovigerous females (excluding the 3 females with premature first instars) from Waikīkī, and 10% of the ovigerous females from French Frigate Shoals. No eyes were visible in the eggs of ovigerous females from Kāne'ohe Bay, but only a total of 9 specimens of the species were collected there.

Cypridina iota, new species

Figs. 18–22

Holotype. BPBM-S 12855, ovigerous female on slides and in vial.

Type Locality. Sta TC-8, 18 Jan 2001, Ala Wai Buoy, Waikīkī, O'ahu.

Paratypes. Waikīkī, Sta TC-8, USNM 1085953, A-1 male on 2 slides and in alcohol.

Distribution. Waikīkī, Sta 8.

Description of adult female (Figures 18, 19). Carapace elongate with convex ventral and dorsal margins; well-developed projecting caudal process with rounded posterior margin (Figure 18*a*). Anterodorsal corner of rostrum evenly rounded, inferior tip of rostrum narrowly rounded, anterior margin of rostrum slightly concave; incisure deep (Figure 18*a,c,d,f*). Outer surface of valve smooth, with 1 minute bristle at midheight of outer margin of rostrum, 4 small bristles forming row ventral to inner end of incisure and setback from anterior edge of valve (Figure 18*b,c,f*), and small bristles along ventral edge of valve. Anterior edge of valve just ventral to incisure with 4 indistinct broad triangular bristles (whether the bristles are on outer or inner side of valve difficult to resolve with certainty) (Figure 18,*c,d,f*). Anteroventral edge of each valve with 8 to 10 bristles; edge of valve in vicinity of the anterior bristle fairly smooth (Figure 18*b,c,f*).

Infold: Narrow list with anterior end on anteroventral infold extends along ventral margin, then broadens to form broad shelf at anterior end of caudal process (Figure 18*c–g*). Rostral infold with about 14 bristles (many divided) plus pair of bristles at inner end of incisure and 2 small bristles proximal and posterior to pair of bristles (Figure 18 *d,f*); about 20 short bristles along anteroventral and ventral list. Posterior edge of ridge forming anterior of pocket of caudal process of right valve with 17 stout spines (Figure 18*e*); posterior edge of ridge of left valve with 15 stout spines (Figure 18*g*); 8 minute bristles form row anterior to posterior edge of ridge. Pocket of caudal process of right valve with 17 stout bristles, some beneath broad anterior ridge (Figure 18*e*). Pocket of caudal process of left valve with 11 stout bristles, some beneath broad anterior ridge (Figure 18*g*). Both valves with minute bristles present just within edge of caudal processes; few minute bristles or pores present in pocket.

Selvage: Lamellar prolongation with smooth outer edge present along anterior and ventral margins, terminating posteriorly near ventral end of caudal process. Prolongation widest and with narrow striations in vicinity of incisure and divides at inner edge of incisure (Figure 18*d,f*).

Central Adductor Muscle Attachments (Figure 19*a*): Consisting of about 10 round and elliptical scars.

Carapace Size (length, height in mm): BPBM-S 12855, 1.37, 0.84.

First Antenna: 1st article bare (Figure 19*e*). 2nd article with ventral, dorsal, and medial spines. 3rd article short, with 2 short bare bristles (1 ventral, 1 dorsal). 4th article with 2 short terminal bristles (1 bare ventral, 1 spinous dorsal). Sensory bristle of short 5th article with 6 long filaments followed by 2 shorter more slender filaments (proximal of these close to base of distal long filament). 6th article with short bare medial bristle. 7th article: a-bristle bare, slightly stouter than bristle of 6th article; bare b-bristle shorter than c- and d-bristles, with proximal rings; c-bristle about twice length of sensory bristle of 5th article, with 7 marginal filaments. 8th article: d- and e-bristles about same length as sensory bristle of 5th article, bare (e-bristle shorter than d-bristle); f-bristle long with 8

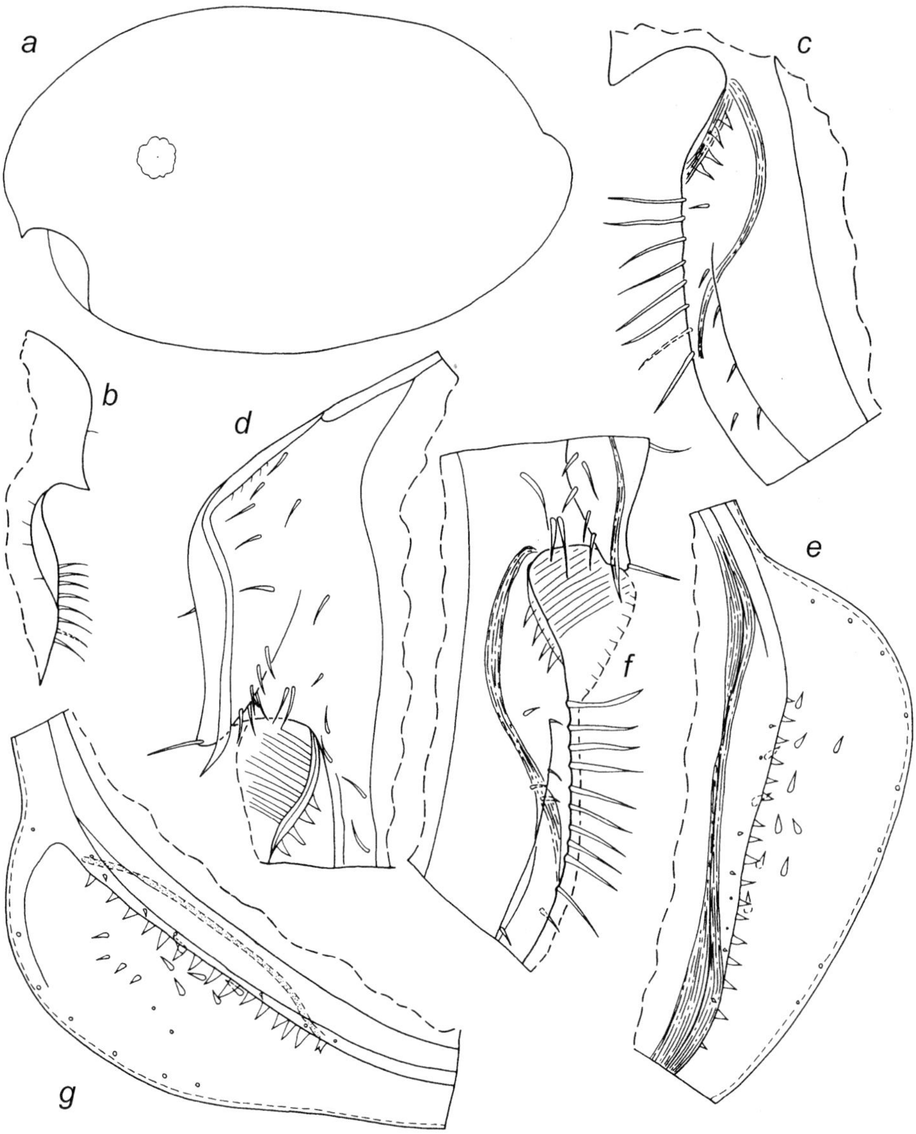

Figure 18. *Cypridina iota,* new species, holotype, BPBM-S 12855, adult female: *a,* complete specimen from left side, length 1.37 mm. Anterior right valve: *b*, ov; *c, d,* iv. *e,* posterior right valve, iv. Left valve, iv: *f,* anterior; *g,* caudal process.

marginal filaments, some with marginal spines; g-bristle longer than f-bristle, with about 11 marginal filaments.

Second Antenna: Protopod with small medial bristle. Endopod with single article with 5 or 6 bristles (1 long filamentlike, 1 medium ringed, 3 or 4 short ringed). Exopod: article 1 with few spines along concave dorsal margin; 2nd article with bristle reaching 7th article, with 9 to 11 ventral spines; articles 3 to 8 with long bristle with natatory hairs but no spines; 9th article with 3 bristles: short dor-

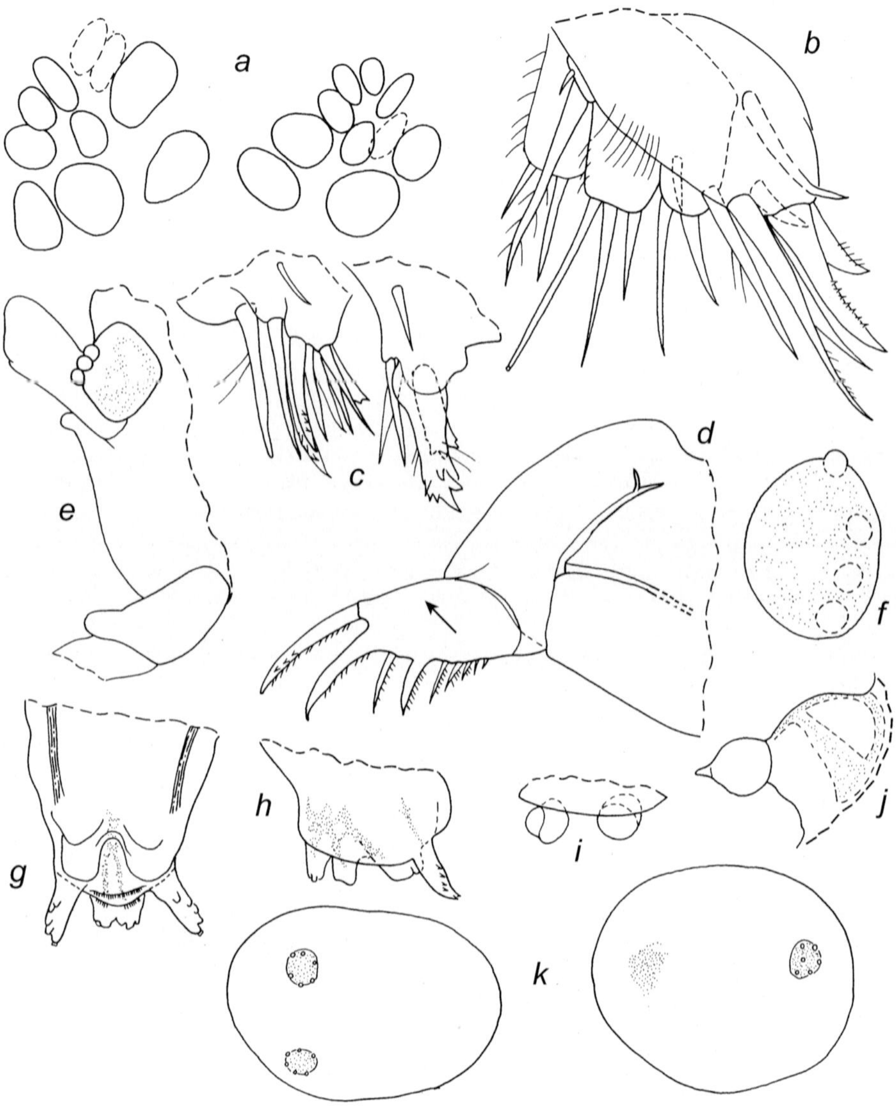

Figure 19. *Cypridina iota,* new species, holotype, BPBM-S 12855, adult female: *a,* central adductor muscle attachments left and right valves, iv; *b,* distal right 5th limb, av; *c,* endites I (on right) and II, left 5th limb, av (nabs); *d,* posterior of body showing right furcal lamella, mv; *e,* anterior of body showing left lateral eye, 1st article of left 1st antenna, and proximal part of left mandible; *f,* lateral eye (not all ommatidia shown); *g, h,* posterior and lateral views of upper lip; *i,* genitalia; *j,* medial eye and Bellonci Organ; *k,* 2 eggs showing lateral eyes.

sal bristle with few short hairs; 2 long bristles with natatory hairs but no spines. No minute spines forming row observed along distal margins of any articles of exopod. Articles 3 to 8 with slender basal spines; spines longer on distal articles; spine on 8th article almost as long as 9th article; 9th article with pointed lateral spine slightly longer than 9th article.

Mandible (Figure 19*e*): Coxa: endite spinous with 2 stout branching terminal spines with smaller triangular process between them; small bristle at base of endite. Basis: dorsal margin with few proximal spines and 3 bristles (1 distal to midlength, 2 subterminal); proximal ventral corner with 2 small medial a-bristles; ventral margin with 1 small b-bristle, 2 c-bristles (1 long, 1 short), and 2 spinous d-bristles (proximal short, distal long). Exopod about same length as dorsal margin of 1st endopod article, with 2 subterminal bristles (distal shorter), and hirsute tip. Endopod: 1st article with 4 ventral bristles (1 long, 2 short, 1 minute). 2nd article: ventral margin with spines and 4 spinelike bristles (2 single, 2 terminal paired); dorsal margin with few proximal spines, 5 short spinous bristles, 3 medium length bristles, and 4 long bristles. 3rd article with 3 claws of equal length (ventral and middle claw with proximal ventral teeth) and 4 ringed bristles (3 near ventral margin (long ventral bristle S-shaped in proximal part and with proximal ventral spines, others short medial with small spines) and 1 lateral bristle near dorsal margin).

Maxilla: Endite I with 6 or 7 bristles and claws; endite II with 5 bristles; endite III with 5 bristles. Precoxa with fringe of dorsal hairs. Coxa with dorsal hairs (no dorsal bristle on BPBM-S 12855; probably broken off during dissection). Basis with 1 short bristle. Exopod with hairs along outer margin, 2 terminal bristles and 1 shorter subterminal bristle (subterminal and longer terminal bristle with long spines). Endopod: 1st article with dorsal spines and 1 alpha-bristle, 2 beta-bristles and 2-pronged cutting tooth. 2nd article with 3 ringed lateral a-bristles, 2 b-bristles, 2 c-bristles, and 3 pectinate clawlike d-bristles.

Fifth Limb: Coxa endite I with 5 bristles (Figure 19*c*); endite II bristles difficult to resolve (about 7 bristles) (Figure 19*c*); endite III with 6 bristles and 1 triangular process. Basis: anterior side with 3 bristles (1 bristle posterior to 2 bristles forming row); basis endite I (main tooth) with 6 stout pectinate cusps, proximal triangular peg, and short spinous proximal bristle; endite II with 3 a-bristles, 3 b-bristles, and 1 ringed, spinous, posterior, proximal c-bristle; anterior side of endite II with long hirsute d-bristle adjacent to smaller bristle (Figure 19*b*). Endopod (Figure 19*b*): 1st article with 1 proximal bristle and 2 terminal bristles (shorter terminal bristle without rings); fused articles 2 and 3 with 3 terminal bristles. Exopod with marginal hairs and 2 terminal bristles (Figure 19*b*).

Sixth Limb: Epipod with 1 or 2 short bare bristles. Endite I with 1 long and 2 short bristles; endite II with 1 or 2 long and 2 short bristles; endite III with 2 long and 1 short bristle; endite IV with 1 long and 1 short bristle. End article with 5 or 6 short and 2 long bristles, stiff spines along lateral edge, and abundant hairs on medial side.

Seventh Limb: 2 or 3 short proximal bristles (1 or 2 on each side), each with 3 bells; edge of terminal segment with 6 bristles, 3 on each side (2 short bristles on each side with 3 bells, 1 long bristle on each side with 4 bells). Comb with total of 6 teeth (middle tooth long, 2 on each side slightly shorter; 1 on each end shorter). Indistinct minute triangular process opposite comb (process could be interpreted to be irregularity in flat edge).

Furca (Figure 19*d*): Each lamella with 9 claws with teeth along posterior edges; claws 2 and 4 nonarticulated; claw 3 narrower than claw 4, but about same length; claw 1 with stout medial teeth and with few distal anterior spines; right lamellar anterior to left by width of base of claw 1.

Bellonci Organ (Figure 19*j*): Short, broad, with triangular tip.

Eyes: Medial eye bare, with brownish coloration in parts (Figure 19*j*). Lateral eye larger than medial eye, with dense black pigment and many obscured ommatidia (Figure 18a, 19*e,f*).

Upper Lip (Figure 19*g,h*): With 2 unpaired anterior processes and 2 paired posterior processes.

Anterior of Body (Figure 19*e*): Small rounded process ventral to base of 1st antenna.

Posterior of Body (Figure 19*d*): Evenly rounded, bare.

Y-Sclerite (Figure 19*d*): With distal ventral branch.

Genitalia (Figure 19*i*): Comprising 2 ovoid processes anterior to furca.

*Brush*like *Bristles*: Not observed.

Eggs: BPBM-S 12855 with 12 oval eggs in marsupium, without appendages, each with 2 black eyes with about 6 visible amber-colored ommatidia. Dimensions of 2 eggs (in mm): length 0.323, width 0.278 (Figure 19*k*, right); length 0.323, width 0. 230 (Figure 19*k*, left).

Gut Content: Gut of BPBM-S 12855 with animal fragments, possibly crustaceans.

Description of A-1 male (Figures 20–22). Carapace elongate with convex ventral and dorsal margins; well-developed projecting caudal process with rounded posterior margin (Figure 20*a*).

Anterodorsal corner of rostrum evenly rounded, inferior tip of rostrum pointed, anterior margin of rostrum slightly concave; incisure deep (Figure 20*a,c,d)*. Outer surface of valve smooth, with 1 minute bristle at midheight of outer margin of rostrum (Figure 20*a,d*), and small bristles along ventral edge of valve. Anteroventral edge of each valve with 9 bristles; edge of valve in vicinity of the anterior bristle fairly smooth (Figure 20*c,d*).

Infold: Narrow list with anterior end on anteroventral infold extends along ventral margin, then broadens to form broad shelf at anterior end of caudal process (Figure 20*d–f*). Rostral infold with about 6 bristles (obscured) plus pair of bristles at inner end of incisure (Figure 20*d,f*); short bristles along anteroventral and ventral list. Posterior edge of ridge forming anterior of pocket of caudal process of right valve with about 20 stout spines (Figure 20*e*); posterior edge of ridge of left valve with 16 stout spines (Figure 20*f*); minute bristles form row anterior to posterior edge of ridge. Pocket of caudal process of right valve with 10 stout bristles, some beneath broad anterior ridge (Figure 20*e*). Pocket of caudal process of left valve with 13 stout bristles, some beneath broad anterior ridge (Figure 20*f*). Both valves with minute bristles present just within edge of caudal processes; few minute bristles or pores present in pocket.

Selvage (Figure 20*c,d*): Lamellar prolongation with smooth outer edge present along anterior and ventral margins, terminating posteriorly near ventral end of caudal process. Prolongation widest and with narrow striations in vicinity of incisure and divides at inner edge of incisure.

Central Adductor Muscle Attachments (Figure 20*b*): Consisting of about 12 round and elliptical scars.

Carapace Size (length, height in mm): USNM 1085953, 1.29, 0.71.

First Antenna (Figure 20*g,h*): 1st article bare. 2nd article with dorsal and medial spines. 3rd article short, with 2 short bare bristles (1 ventral, 1 dorsal). 4th article with 2 short terminal bristles (1 ventral, 1 dorsal). Sensory bristle of short 5th article with 1 short proximal filament, 6 long filaments, and 2 shorter more slender filaments (proximal of these close to base of distal long filament). 6th article with short bare medial bristle. 7th article: a-bristle bare, slightly stouter than bristle of 6th article; b-bristle shorter than c- and d-bristles, with proximal rings and 4 short proximal spines; c-bristle about twice length of sensory bristle of 5th article, with 8 marginal filaments. 8th article: d- and e-bristles about same length as sensory bristle of 5th article, bare (e-bristle shorter than d-bristle); f-bristle long, tip missing) with 6 marginal filaments present, some with marginal spines; g-bristle longer than f-bristle, with about 9 marginal filaments.

Second Antenna: Protopod with small medial bristle. Endopod with single article with 4 bristles (1 long filamentlike, 1 medium ringed, 2 short ringed) (Figure 21*a,b*).

Exopod: article 1 with few spines along concave dorsal margin; 2nd article with bristle reaching 7th article, with 9 to 11 ventral spines; articles 3 to 8 with long bristle with natatory hairs but no spines; 9th article with 3 bristles: short dorsal bristle with few short hairs; 2 long bristle with natatory hairs but no spines. No minute spines forming row observed along distal margins of any articles of exopod. Articles 3 to 8 with slender basal spines; spines longer on distal articles; spine on 8th article almost as long as 9th article; 9th article with pointed lateral spine slightly shorter than 9th article.

Mandible (Figure 21*c*): Coxa: endite spinous with 2 stout branching terminal spines with smaller triangular process between them; small bristle at base of endite. Basis: dorsal margin with few proximal spines and 3 bristles (1 distal to midlength, 2 subterminal); proximal ventral corner with 2 small medial a-bristles; ventral margin with 1 small b-bristle, 2 c-bristles (1 long, 1 short), and 2 spinous d-bristles (proximal short, distal long). Exopod about same length as dorsal margin of 1st endopod article, with 2 subterminal bristles (distal shorter), and hirsute tip. Endopod: 1st article with 4 ventral bristles (1 long, 2 short, 1 minute). 2nd article: ventral margin with spines and 4 spinelike bristles (2 single, 2 terminal paired); dorsal margin with few proximal spines, 1 short spinous bristle, 2 medium length bristles, and 4 long bristles. 3rd article with 3 claws of equal length (ventral and middle claw with proximal ventral teeth) and 4 ringed bristles (3 near ventral margin (long ventral bristle S-shaped in proximal part and with proximal ventral spines, others short medial with small spines) and 1 lateral bristle near dorsal margin).

Maxilla: Endite I with 7 bristles and claws; endite II with 5 bristles; endite III with 5 bristles (Figure 21*d*). Precoxa with fringe of dorsal hairs. Basis with short ventral bristle and long dorsal

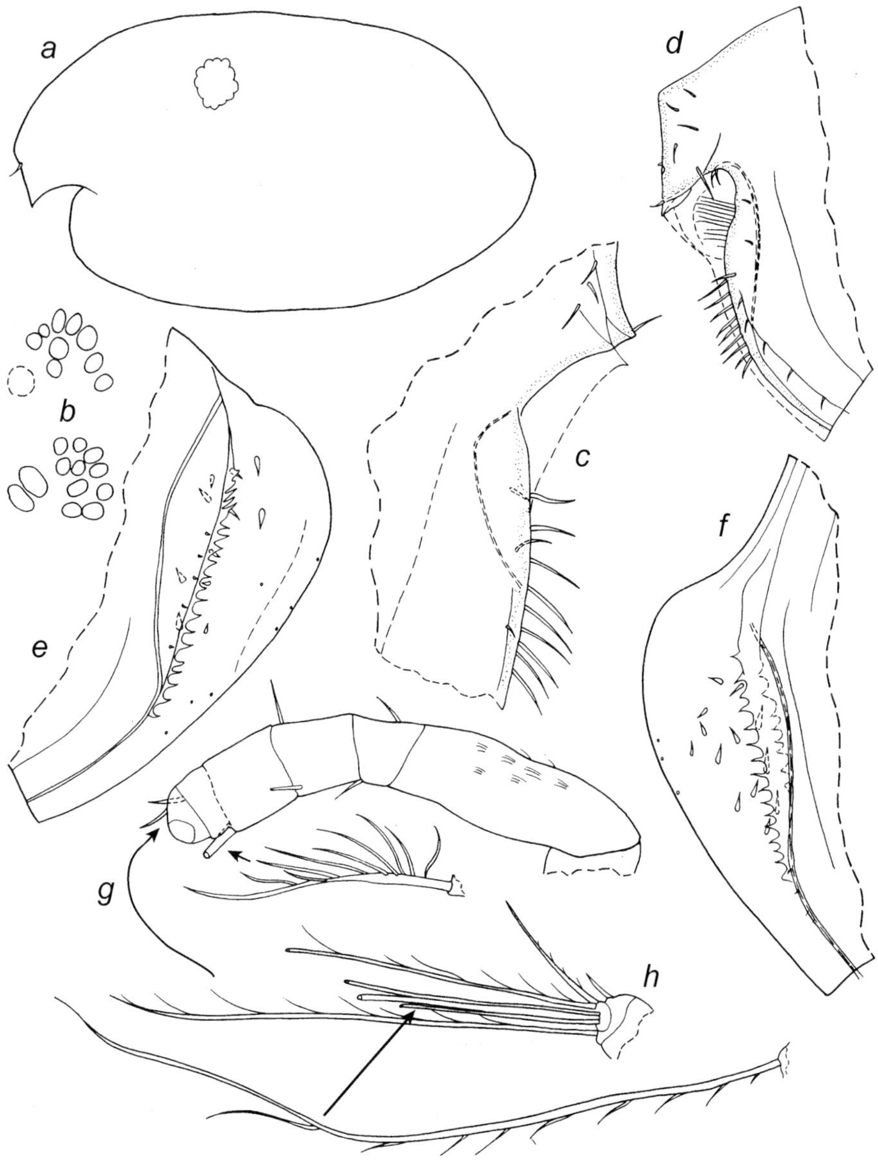

Figure 20. *Cypridina iota,* new species, paratype, USNM 1085953, A-1 male: *a,* complete specimen, length 1.29 mm; *b,* central adductor muscle attachments of left (top) and right valves, iv (not all muscle attachments shown). Anterior right valve: *c,* ov; *d,* iv. Caudal process, iv: *e,* right valve; *f,* left valve; *g,h,* left 1st antenna, lv (nabs).

bristle. Exopod with 1 proximal and 2 terminal bristles (not shown). Endopod (Figure 21*e*): 1st article with dorsal spines and 1 alpha-bristle (beta-bristles and cutting tooth obscured). 2nd article with 3 ringed lateral a-bristles, 5 ringed bristles, and 3 pectinate clawlike d-bristles.

Fifth Limb: Bristles of coxa endites I to III difficult to resolve (some bristles shown in Figure 22*b,c*). Basis: anterior side with 4 bristles (1 bristle lateral to 3 bristles forming row) (Figure 22*a*); basis endite I (main tooth) with 5 stout pectinate cusps, proximal triangular peg, and short spinous proximal bristle (Figure 21*f*); endite II with 3 a-bristles, 4 b-bristles, and 1 ringed, spinous, posteri-

Figure 21. *Cypridina iota,* new species, paratype, USNM 1085953, A-1 male: *a,* protopod and endopod, right 2nd antenna, mv; *b,* protopod and endopod left 2nd antenna, mv; *c,* left mandible, mv; *d,* endites left maxilla, lv; *e,* tip left maxilla, lv; *f,* distal right 5th limb, pv (nabs); *g,* distal left 5th limb, av (nabs).

or, proximal c-bristle; anterior side of endite II with long hirsute d-bristle (Figure 21*f,g*). Endopod (Figure 21*f,g*): 1st article with 1 proximal bristle and 2 terminal bristles (shorter terminal bristle without rings); fused articles 2 and 3 with 3 terminal bristles. Exopod with marginal hairs and 2 terminal bristles (Figure 21*f,g*).

Sixth Limb (Figure 22*d*): Epipod with 2 short bare bristles. Endite I with 1 long and 2 short bristles; endite II with 2 long and 2 short bristles; endite III with 2 long and 1 short bristle; endite IV

Figure 22. *Cypridina iota,* new species, paratype, USNM 1085953, A-1 male: *a,* large claw of basis endite I and basis left 5th limb, av (nabs); *b,* coxa endites I–III left 5th limb, av (nabs); *c,* coxa endites I–III right 5th limb, pv (nabs); *d,* right 6th limb, lv; *e, f,* 7th limbs; *g,* left furcal lamella, lv; *h,* posterior of body from left side showing copulatory organ, anus (dashed), Y-Sclerite and girdle; *i,* lateral eye.

with 1 long and 1 short bristle. End article with 6 short and 2 long bristles, stiff spines along lateral edge, and abundant hairs on medial side.

Seventh Limb (Figure 22*e,f*): 3 short proximal bristles (1 or 2 on each side), each bare or with few small terminal spines; edge of terminal segment with 5 or 6 bristles, 2 or 3 on each side (3 or 4 short bristles bare or with few small terminal spines, 1 long bristle on each side with 2 bells). Comb with total of 5 teeth (middle tooth long, 2 on each side shorter); short peg at inner side of middle tooth (detail in Figure 22*f*).

Furca (Figure 22*g,h*): Each lamella with 8 claws with teeth along posterior edges; claws 2 and 4 nonarticulated; claw 3 narrower than claw 4, but slightly longer; claw 1 with stout medial teeth and with few distal anterior spines; right lamellar anterior to left by width of base of claw 1.

Bellonci Organ: Obscured.

Eyes: Medial eye with brown pigment. Lateral eye with dense black pigment and many obscured ommatidia (Figures 20*a*, 22*i*).

Upper Lip: With 2 unpaired anterior processes and 2 paired posterior processes, similar to that of adult female.

Anterior of Body: Obscured.

Posterior of Body (Figure 22*h*): Bare, with scalloped edge dorsal to posterior end of Y-Sclerite.

Y-Sclerite (Figure 22*h*): Typical for subfamily.

Genitalia (Figure 22*h*): Poorly developed processes anterior to furca may represent copulatory organ.

*Brush*like *Bristles*: None observed.

Comparisons. The carapace of *C. iota* differs from that of *C. alpha* mainly in having spines along the anterior ridge of the caudal process in both the left and right valves. *C. alpha* has them only in the right valve. The caudal process of the left valve of *C. iota* also bears stout bristles absent in the left valve of *C. alpha.* The 6th limbs of the single female of *C. iota* bear 1 or 2 epipod bristles and a 6th limb of the single male A-1 instar of *C. iota* bears 2 epipod bristles; whereas, the 6th limbs of 2 adult females of *C. alpha* examined have 4 epipod bristles. This suggests that the 6th limb of *C. iota* bears fewer epipod bristles than the limb of *C. alpha*: 1 or 2 compared to 4, but additional specimens should be examined to determine variability. The valves of *C. iota* differ from those of *C. spinula* in having 8 to 10 bristles along the anteroventral edge of each valve and in not having many stout bristles in the pocket of the caudal process of the right valve. The carapace of *C. melletini* Kornicker & Harrison-Nelson differs from that of *C. iota* in not having spines along the anterior ridge of the caudal process of the left valve as well as not having many stout bristles in the pocket of the caudal process of the right valve.

Remarks. The senior author is inclined to believe that the presence of spines on the inner ridge of the caudal process of the left valve of *C. iota,* which is the main difference between *C. iota* and *C. alpha,* is either a polymorphism or a recurrent mutation of *C. alpha,* but because so great a morphological difference would be unusual according to my present understanding of intraspecific variation within myodocopids, *C. iota* has been described herein as a new species. The 2 specimens of *C. iota* in the collection are from the same sample, which also contains *C. alpha.* Many samples contain only *C. alpha*. It seems unlikely that if the morphological difference peculiar to *C. iota* is a variation within *C. alpha,* both specimens of *C. iota* would be collected in the same sample, but it could be a coincidence.

Discussion. One of the more difficult decisions facing a taxonomist is in discriminating 2 sympatric species having few character differences and without intermediates. Ross (1974: 131) suggested the following method: "If sets 1 and 2 differ in 2 or more character states, they are probably 2 species. The larger the number of differences, the greater is this probability. If the difference is in only one character state, 2 species might be involved, but there is a possibility that the differences represent 2 alleles, one exhibiting complete dominance." There is no hard and fast methodology. The problem seemed to occur more often in the collections of Myodocopina from Hawai'i than previously encountered in other areas by the senior author. Hopefully, future studies by others will establish more firmly whether or not the discriminations made herein are correct.

Superfamily SARSIELLOIDEA Brady & Norman

Composition and distribution. The Sarsielloidea contain the families Sarsiellidae Brady & Norman, Rutidermatidae Brady & Norman, Philomedidae Müller, and Pseudophilomedidae Kornicker. Circumglobal distribution.

Family PSEUDOPHILOMEDIDAE Kornicker

The Pseudophilomedidae include the subfamily Pseudophilomedinae Kornicker.

Subfamily PSEUDOPHILOMEDINAE Kornicker

Composition and distribution. This family includes 5 genera of which only *Harbansus* is represented in the present collection. Circumglobal at subtidal to abyssal depth.

Genus *Harbansus* Kornicker

Harbansus Kornicker, 1978: 13. **Type species**. *Harbansus bradmyersi* Kornicker, 1978, by original designation.

Composition and distribution. Including the new species described herein, this genus contains 20 species plus 2 left in open nomenclature (Appendix 2). Cosmopolitan at depths of subtidal to 1015 m (Kornicker, 1978: 13; 1992: 97; 1995: 19). The type species is known only from the continental shelf off Oxnard, California, depth 18.3 m (Kornicker, 1978: 24). Hartmann (1984: 137) incorrectly listed the species (as *Habansus bradmeyersi*) as being found off the Great Barrier Reef, Hawai'i. *H. barnardi* Kornicker was described from Kāne'ohe Bay, O'ahu, by Kornicker (1978: 45) and a new species is described herein from French Frigate Shoals.

Relationship between lateral eyes of *Harbansus* and water depth. The relationship of lateral eyes and water depth of *Harbansus* was discussed by Kornicker (1992: 108). The additional species of the genus described since then support a decrease in the number of species with lateral eyes with water depth (Table 3). The adult males of 6 species of *Harbansus* are now known: 5 have lateral eyes; 4 of those have females with eyes and one (*H. dayi*) has females without eyes. The single male without eyes belongs to *H. boweni,* which is also without eyes on the female.

Methods of locomotion by species of *Harbansus*. The bristles of adult females and juveniles of some species of *Harbansus* have natatory hairs on bristles of the exopod of the 2nd antenna, indicating that those species are swimmers; whereas other species have shorter bristles with only spines, indicating that those species are crawlers (Kornicker, 1978: 11, 1992: 105). Kornicker (1992: 105), noted that both male and female juveniles then known have exopod bristles similar to those of the adult female, and therefore, the type of bristles on juveniles can be used to estimate the type of bristles on adult females. The new species *H. hox,* described here is an exception in that the instar III female is a crawler whereas the instar IV female and male and adult female are swimmers. Adult

Table 3. Distribution of lateral eyes of adult females of species of *Harbansus* with water depth (based on data in Appendix 2).

Depth (m)	Number of species (with and without eyes)	Species with eyes (%)
0–100	11	82
101–400	9	44
401–1015	3	0

Table 4. Distribution of swimming and crawling females (either instars III, IV or adult) of species of *Harbansus* with water depth (based on data in Appendix 2). Species occupying more than one depth range are included in each range.)

Water depth (m)	Number of species (swimmers + crawlers)	Swimming species (%)	Crawling species (%)
0–100	12	42	58
101–400	10	20	80
401–1015	4	0	100

Table 5. Relationship of method of locomotion (swimming or crawling) of females (either instars III, IV, or adult) of species of *Harbansus* with ability to see (based on data in Appendix 2).

Locomotion (# of species)	Species with lateral eyes (%)	Species without lateral eyes (%)
Swimmers (6)	100	0
Crawlers (19)	50	50

males are known for only 6 species of *Harbansus*. All are swimmers regardless of whether the females or juveniles are swimmers or crawlers. The number of species with females capable of swimming appears to decrease with water depth (Table 4). Swimming females are more likely to have eyes than crawling females, but some species of both types have eyes (Table 5). Exopodial bristles of swimmers have stout proximal sections consisting of long "segments", each with a stout ventral spine, and a slender distal part with closely spaced rings with natatory hairs. Exopodial bristles of crawlers consist entirely of long "segments" with spines and are similar to the proximal section of bristles of swimmers. Examination of the tips of bristles of crawlers show them to terminate with a spine, and the tips show no indication that a slender part may have broken off. The following species have crawling adult or instar IV females: *H. schornikovi, H. bradmyersi, H. rhabdion, H. mayeri, H. dayi, H. bowenae, H. barnardi, H. magnus, H. vix, H. ferox, H. felix, H. tenax, H. hapax, H. vortex, H.* species A, *H.* species B. The following species have swimming adult or instar IV females: *H. paucichelata, H. slatteryi, H. thrix, H. vatrax, H. flax, H. hox.*

Harbansus hox, new species

Figs. 23–29

Holotype. BPBM-S 12868, ovigerous female in alcohol.

Type Locality. Sta TC-13, French Frigate Shoals, ovigerous female in alcohol.

Paratypes. French Frigate Shoals: Sta TC-8, USNM 1085954, ovigerous female on slide and 1 egg in alcohol (valves lost). Sta TC-13, USNM 1085955, 1 adult female without eggs. BPBM-S 12849, 1 instar IV female; USNM 1085956, 1 instar IV male; BPBM-S 12848, 1 instar III female (valves lost); USNM 1085950, 1 instar IV female (valves lost)

Remarks. Two additional ovigerous females from Sta TC-13 were lost.

Distribution. French Frigate Shoals, Sta TC-8, TC-13.

Description of adult female (Figures 23–25). Each valve with overhanging rostrum, well-developed caudal process, and numerous large shallow fossae (Figure 23*a*). Anterior edge and anteroventral edge of rostrum with 10 small tubercles (Figure 23*a,d*). Rostrum with broad low triangular lateral rib. Posterodorsal corner with small process bearing single bristle. Surface and valve edges with long single bristles.

Infold: Rostral infold with row of 6 bristles, and a pair of bristles at inner edge of incisure (Figure 23*d*). Anteroventral infold with 1 bristle near incisure and about 6 narrow ridges (Figure 23*d*). Infold of caudal process with 7 frondlike bristles along ridge forming pocket (Figure 23*e*). Minute bristle at posterior tip of caudal process. Additional bristle on inner edge of infold anterior to caudal process (Figure 23*e*). Posteroventral infold anterior to caudal process with several bristles forming row parallel to edge of valve.

Ornamentation (Figure 23*a*): Small processes at posterodorsal corner of valve; Valve surface with abundant well defined small fossae (representative fossae shown in Figure 23*a*). Surface and edges of valves with single bristles.

Central Adductor Muscle Attachments: Numerous elongate and ovoid muscle scars (Figure 23*b,c*).

Selvage (Figure 23*d*): Divided at inner end of incisure; with marginal fringe.

Carapace Size (length, height in mm): USNM 1085954, 0.97, 0.50 (1 egg). USNM 1085955: 1.01. 0.52 (no eggs). Two specimens from TC-13 (lost): 0.97, 0.53 (3 eggs); 0.95, 0.51 (2 eggs); BPBM-S 12868, 0.97, 0.54 (1 egg).

First Antenna (Figure 23*f*): 1st article bare. 2nd article with spines along ventral and dorsal margins and on medial surface, and with 1 spinous dorsal bristle. 3rd article short with spines along ventral and dorsal margins and 3 bristles (1 ventral, 2 dorsal). 4th article with indistinct ventral and dorsal spines and 4 or 5 bristles (2 or 3 ventral, 2 dorsal). Sensory bristle of 5th article with 1 short and 3 longer subterminal filaments excluding stem; tip of stem with 2 indistinct minute spines. Bristle of minute 6th article medial and about same length as 5th article. 7th article: a-bristle longer than bristle of 6th article; b-bristle shorter than c-bristle, with 2 indistinct short filaments near midlength; c-bristle long with 1 short proximal filament, 4 subterminal filaments, and stem with 2 spines at tip. 8th article: d- and e-bristles long bare; f-bristle with 1 short proximal filament and 3 subterminal filaments; g-bristle long with 3 subterminal filaments. (Filaments of bristles of 7th and 8th articles indistinct and number given approximate.)

Second Antenna: Protopod bare. Endopod with 2 articles (Figure 23*g*): article 1 with 2 small proximal ventral bristles; article 2 with 1 long spinous ventral bristle near midlength and 1 minute terminal process. Exopod with 9 articles (Figure 24*a*): article 2 with ventral bristle reaching past 9th article with 10 stout ventral spines and smaller ventral spines on short narrower tip, and small recurved terminal medial spine; article 3 with long bristles with 8 stout ventral spines along stout proximal half and natatory hairs on narrower distal half; articles 4 to 8 with bristles similar to that of article 2; article 9 small with 2 bristles (ventral bristle about one-half length of bristle of 8th article, with short slender spines; dorsal bristle short bare). Articles 2 to 8 with spines forming row along medial distal margin (spines stouter at dorsal corner); articles 4 to 8 with small basal spines; article 9 with minute medial spine near base of dorsal bristle. Pivot sclerite slightly concave dorsally.

Mandible: (Figure 24*d*): Coxa endite bifurcate, spinous, with indistinct minute bristle near base. Basis: medial side spinous and with 3 small bristles in proximal ventral corner, 1 short bristle near midlength with base close to ventral margin; ventral margin with 1 long spinous subterminal bristle; lateral surface with 2 bristles near midlength and fairly close to ventral margin; dorsal margin with 1 bristle near midlength and 2 terminal bristles. Exopod hirsute, about three-fourths length of 1st endopod article, with long subterminal bristle and short terminal bristle. Endopod: 1st article with 2 long and 2 short subterminal ventral bristles. 2nd article: ventral margin with bristles forming 2 distal groups (proximal group with 2 bristles, distal group with 3 bristles including 1 clawlike bristle); dorsal margin with bristles forming 2 groups near midlength (2 bristles in proximal group, 6 in distal group). 3rd article with 2 stout claws, 1 minute dorsal claw, and 3 longer bristles.

Figure 23. *Harbansus hox*, new species, paratype, USNM 1085954, adult female: *a,* complete specimen from right side, length 0.97 mm; *b,* detail of central adductor muscle attachments from "*a*"; *c,* central adductor muscle attachments right valve, iv; *d,* anterior left valve, iv; *e,* posterior left valve, iv; *f,* left 1st antenna, mv; *g,* endopod left 2nd antenna, mv; *h,* left maxilla, mv (nabs); *i,* tip left maxilla, mv; *j,* right lateral eye.

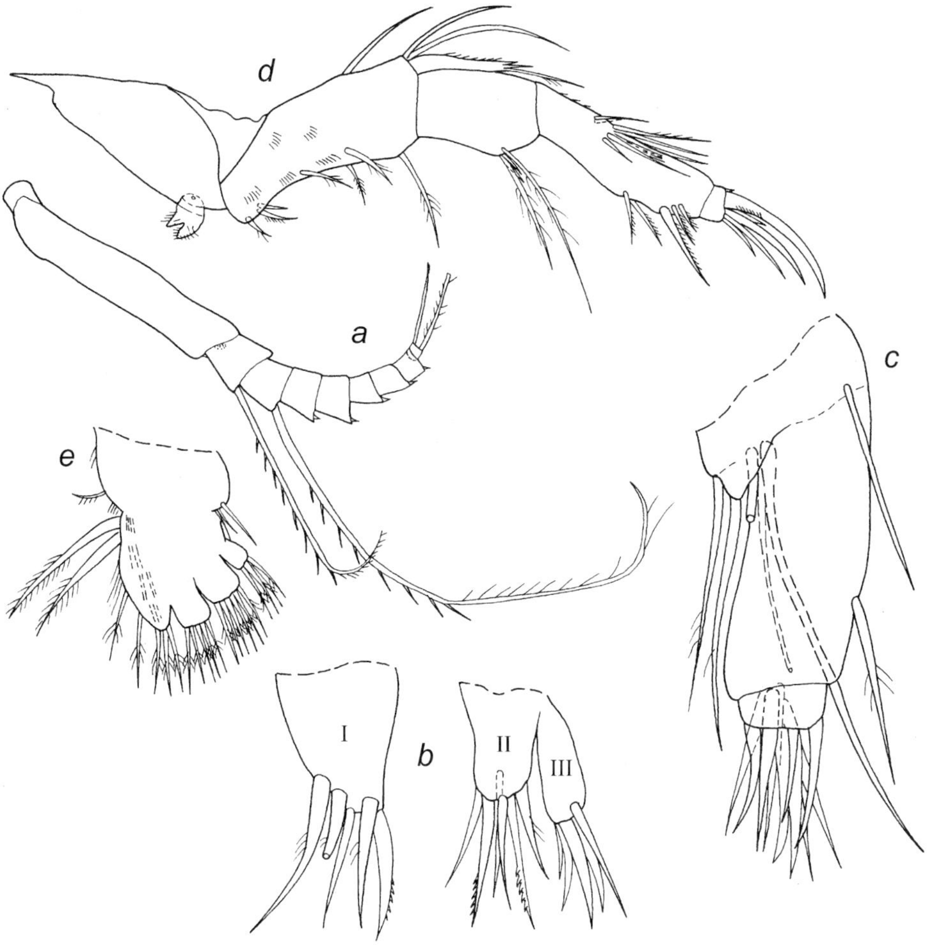

Figure 24. *Harbansus hox*, new species, paratype, USNM 1085954, adult female: *a,* exopod left 2nd antenna, lv; *b,* endites left maxilla, mv; *c,* distal left maxilla, mv; *d,* right mandible, lv; *e,* right 6th limb, lv.

Maxilla (Figures 23*h,i,* 24*b,c*): Precoxa endite I with 5 bristles. Coxa with spinous dorsal bristle; endite II with 6 bristles, endite III with 5 terminal bristles. Basis with 3 terminal bristles (1 dorsal, 2 ventral). Exopod with 3 bristles (2 long, 1 short). Endopod: 1st article with 1 alpha- and 3-beta bristles. 2nd article with 2 stout claws and about 7 bristles (not all shown).

Fifth Limb (Figure 25*a,b*): Epipod with 39 spinous bristles. Coxa: endites I, II, and III obscured, each with approximately 4 or 5 bristles. Basis: endite I with tooth with 4 prongs, 1 short bristle on edge proximal to prongs, 2 anterior bristles near midwidth along distal edge, and 1 small bristle at outer distal corner; endite II with large squarish tooth; proximal bristles on posterior face obscured. Exopod with 2 spinous bristles. Endopod: 1st article with about 3 bristles; 2nd article with 1 proximal and 3 or 4 terminal bristles.

Sixth Limb (Figure 24*e*): With 1 epipod bristle. Endite I with 3 bristles; endite II with 3 bristles; endite III with 6 bristles; endite IV with 5 bristles. End article not prolonged posteriorly, with 7 bristles (5 anterior spinous bristles separated by space from 2 stout plumose bristles).

Seventh Limb (Figure 25*c*): Each limb with 2 proximal bristles, 1 on each side with 4 bells, and

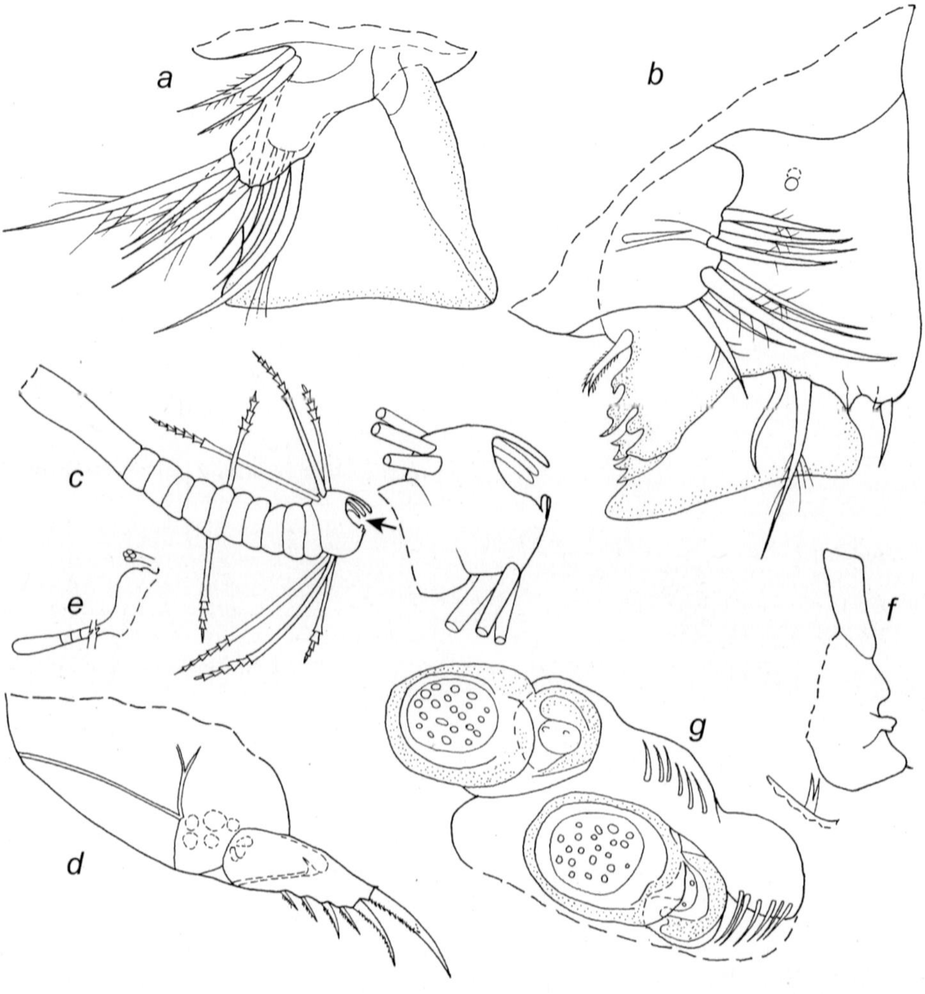

Figure 25. *Harbansus hox*, new species, paratype, USNM 1085954, adult female: *a,* distal right 5th limb, pv; *b,* distal left 5th limb, av; *c*, 7th limb; *d*, posterior of body and right furcal lamella; *e*, left lateral eye, medial eye, and Bellonci Organ; *f*, anterior of body from right side and bifurcate coxa endite of left mandible in place on body; *g*, genitalia with attached spermatophores.

6 terminal bristles, 3 on each side with 4 to 6 bells. Terminus with comb of 7 curved teeth opposite 2 small pegs.

Furca (Figure 25*d*): Each lamella with 6 claws; claws 1, 2, and 4 stout; claws 3, 5, and 6 more slender and shorter; all claws with short teeth along posterior edges; claw 1 with stout teeth proximal to 1 stout medial tooth at midlength. Claw 1 of right lamella anterior to claw 1 of left lamella by about width of base of claw 1.

Bellonci Organ (Figure 25*e*): Broken, elongate with proximal annulations and broadly rounded tip.

Eyes: Lateral eye with 3 ommatidia, some divided (Figures 23*a, j,* 25*e*). Medial eye larger than lateral eye, bare (Figure 25*e*).

Upper Lip (Figure 25*f*): With minute anterior spine.

Anterior of Body (Figure 25*f*): With protuberance between upper lip and 1st antenna.

Posterior of Body (Figure 25*d*): Bare.

Y-Sclerite (Figure 25*d*): With ventral branch.

Genitalia (Figure 25*g*): 2 ovals, each with attached spermatophore. Row of 6 minute bristles adjacent to each oval.

Eggs in Marsupium (Figure 23*a*): USNM 1085954 with 1 egg, length 0.294 mm. BPBM-S 12868 with 1 egg. Two specimens from TC-13 (lost): with 2 and 3 eggs.

Description of instar III female (Figures 26, 27). Carapace shape similar to that of adult female (Figure 26*a–c*).

Infold: Rostral infold with row of 4 or 5 bristles (Figure 26*b*, bristles dashed) and a pair of bristles near inner end of incisure. Anteroventral infold with 1 bristle near incisure and about 7 narrow ridges. Infold of caudal process with 3 or 4 frondlike bristles along ridge forming pocket (Figure 26*c*). Minute bristle at posterior tip of caudal process. Additional small bristle at inner edge of infold anterior to caudal process.

Ornamentation (Figure 26*b*): Similar to that of adult female.

Central Adductor Muscle Attachments (Figure 26*a,d*): Numerous elongate and oval muscle scars.

Selvage: Similar to that of adult female.

Carapace Size (length, height in mm): BPBM-S 12848, 0.63, 0.34.

First Antenna (Figure 26*e*): 1st article bare. 2nd article with spines along dorsal margin and with 1 spinous dorsal bristle. 3rd article short, with spines along dorsal margin and 3 bristles (1 ventral, 2 dorsal). 4th article with spines along ventral and dorsal margins and 2 or 3 bristles (1 ventral, 1 or 2 dorsal). Sensory bristle of 5th article with subterminal filament. 6th article with short medial bristle. 7th article: a-bristle longer than bristle of 6th article; b-bristle shorter than c-bristle, bare; c-bristle long with subterminal filament; d- and e-bristles long, bare; f- and g-bristles long, each with subterminal filament.

Second Antenna: Protopod bare (Figure 26*f, g*). Endopod with 2 articles: article 1 with 1 small ventral bristle; article 2 with long spinous ventral bristle and small indistinct terminal process (Figure 26*g*). Exopod with 9 articles: articles 2 to 8 with short stout bristles with stout spines, but no natatory hairs (Figure 26*h,i*); 9th article with 2 short bristles (dorsal longer and with indistinct small spines; ventral bare). Pivot sclerite slightly curved (Figure 26*f*).

Mandible (Figure 26*j*): Coxa endite bifurcate, spinous. Basis: ventral margin with 3 small bristles in proximal corner, 1 short bristle near midlength, and 1 short subterminal bristle. Lateral surface with 1 small bristle near midlength and close to ventral margin; dorsal margin with 1 bristle distal to midlength and 2 terminal bristles. Exopod about two-thirds length of dorsal margin of 1st endopod article, with 2 terminal bristles. Endopod: 1st article with 2 long and 1 short bristle in distal ventral corner. Article 2: ventral margin with bristles forming 2 distal groups (proximal group with 2 bristles; distal group with 3 bristles including 1 clawlike bristle); dorsal margin with about 7 bristles near midlength. Article 3 with 2 stout claws and about 3 bristles (tip obscured).

Maxilla (Figure 27*a*): Endites I and II each with 4 or 5 bristles; endite III with about 6 bristles. Coxa with 1 dorsal bristle. Basis with 3 terminal bristles (2 ventral, 1 dorsal). Exopod small with 3 bristles (2 long, 1 short). Endopod: article 1 with 1 distal dorsal bristle, 2 alpha bristles and 2 beta bristles; article 2 with 2 stout claws, 1 short claw, and 6 bristles.

Fifth Limb (Figure 27*b–d*): Epipod with 31 spinous bristles (Figure 27*b*). Bristles of coxa endites I to III indistinct (endite III with 5 bristles). Basis (Figure 27*c,d*): endite I elongate with terminal teeth; endite II with large squarish tooth. Exopod with 2 bristles. Endopod: bristles partly obscured (1st article with 3 bristles; endopod II with 1 proximal and 3 terminal bristles).

Sixth Limb (Figure 27*e*): With 1 epipod bristle. Endite I with 2 bristles; endite II with 3 bristles; endite III with 6 or 7 bristles; endite IV with 5 bristles. End article not prolonged posteriorly, with 7 bristles (5 anterior separated by space from 2 stout plumose bristles).

Seventh Limb (Figure 27*f*): Elongate, bare.

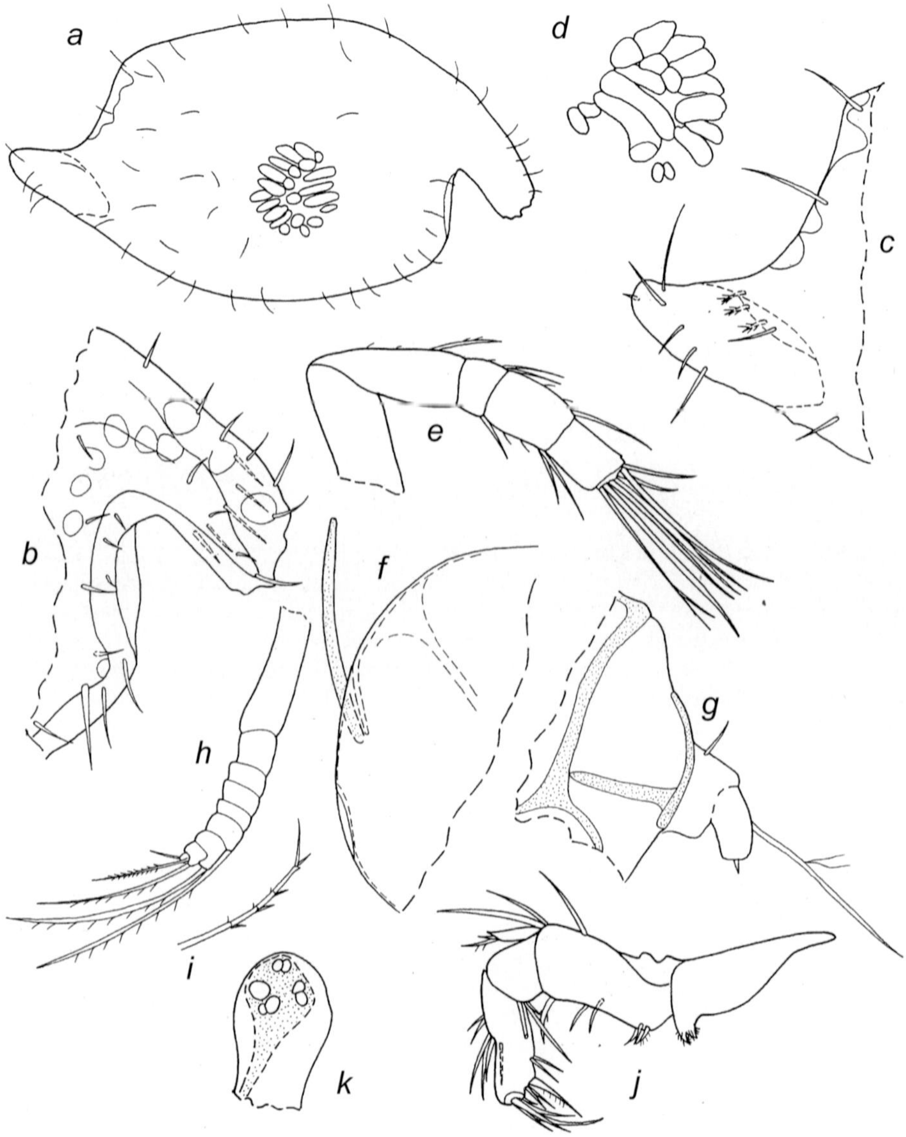

Figure 26. *Harbansus hox*, new species, paratype, BPBM-S 12869, female instar III: *a,* complete specimen from right side, length 0.63 mm; *b, c,* anterior and posterior right valve, ov; *d,* central adductor muscles left valve, ov; *e,* right 1st antenna, lv; *f,* posterior end protopod right 2nd antenna and pivot sclerite (stippled), lv; *g,* endopod left 2nd antenna, mv; *h,* exopod right 2nd antenna, lv (nabs): *i,* bristle of 4th article of exopod of left 2nd antenna, lv; *j,* right mandible, mv; *k,* left lateral eye.

Furca (Figure 27*g*): Similar to that of adult female.

Bellonci Organ: Broken during dissection. Distal half elongate with rounded tip (Figure 27*h*). Proximal part not observed, but presumably with annulations similar to those on adult female organ.

Eyes: Lateral eye with 3 or 4 ommatidia, some divided (Figure 26*k*). Medial eye larger than lateral eye.

Upper Lip: Not observed.

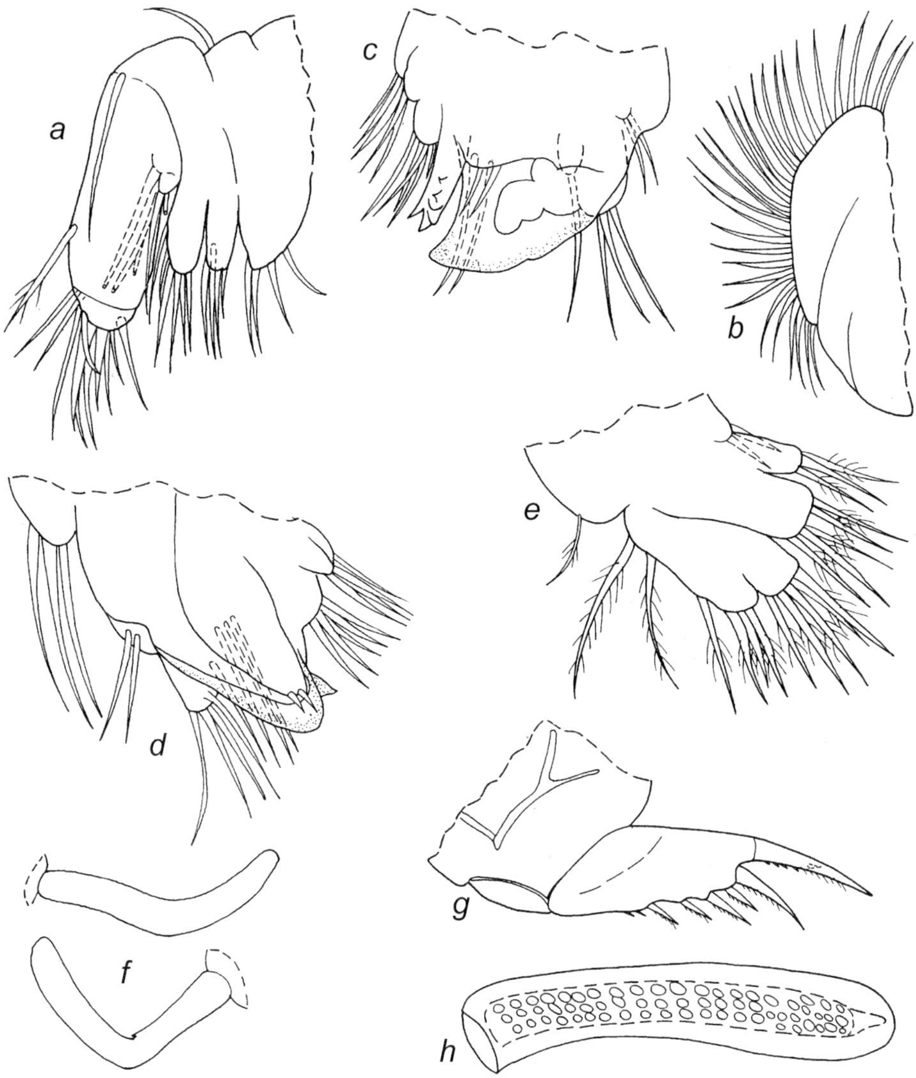

Figure 27. *Harbansus hox*, new species, paratype, BPBM-S 12869, female instar III: *a,* maxilla; *b,* epipod right 5th limb as seen through valve, lv (nabs); *c,* distal right 5th limb, pv; *d,* distal left 5th limb, pv; *e,* right 6th limb, lv; *f,* right and left 7th limbs, lv; *g,* right furcal lamella and Y-Sclerite; *h,* tip of Bellonci Organ.

Posterior of Body (Figure 27*g*): Bare.

Y-Sclerite (Figure 27*g*): Similar to that of adult female.

Genitalia: Absent.

Eggs: Without unextruded eggs.

Description of instar IV female (Figure 28). Carapace shape and ornamentation similar to that of adult female (Figure 28*a*).

Infold: Rostral infold with row of 4 or 5 bristles and a pair of bristles at inner edge of incisure.

Anteroventral infold with 1 bristle near inner end of incisure and about 6 narrow ridges. Infold of caudal process with 6 frondlike bristles along ridge forming pocket and minute bristle anterior to ridge near inner edge of infold. Minute bristle at posterior tip of caudal process. Additional bristles along inner edge of posteroventral infold anterior to caudal process.

Central Adductor Muscle Scars (Figure 28*b*): Numerous elongate and ovoid muscle scars.

Selvage: Similar to that of adult female.

Carapace Size (length, height in mm): BPBM-S 12848, 0.84, 0.44. BPBM-S 12849, 0.76, 0.40.

First Antenna (Figure 28*c*): Except for having 1 instead of 2 dorsal bristles on the 2nd article, limb similar to that of the adult female. Filaments on bristles of articles 5, 7, and 8 not counted.

Second Antenna (Figure 28*d*): Similar to that of adult female. Bristles of exopod long with proximal spines and distal natatory hairs as on adult female.

Mandible (Figure 28*e*): Coxa endite bifurcate, spinous (minute bristle not observed near base). Basis: medial side spinous and with 3 small bristles in proximal ventral corner and 0 or 1 bristle at midlength near ventral margin; lateral surface with 2 bristles near midlength very close to ventral margin; ventral margin with1 long terminal bristle; dorsal margin with 1 bristle near midlength and 2 terminal bristles. Exopod hirsute, about two-thirds length of 1st endopod article, with long subterminal bristle and short terminal bristle. Endopod: 1st article with 3 terminal bristles (2 long, 1 short). 2nd article: ventral margin with bristles forming 2 distal groups (proximal group with 2 bristles, distal group with 3 bristles including 1 long clawlike bristle); dorsal margin with bristles forming 2 groups near midlength (2 bristles in proximal group, 6 in distal group). 3rd article with 2 stout claws, 1 minute dorsal claw, and 3 longer bristles.

Maxilla (Figure 28*f*): In general, similar to that of adult female. (Not all bristles shown on illustrated limb.)

Fifth Limb (Figure 28*g*): Partly obscured; in general, similar to that of adult female.

Sixth Limb (Figure 28*h*): With 1 epipod bristle. Endite I with 2 bristles; endite II with 3 bristles; endite III with 5 bristles; endite IV with 5 bristles. End article not prolonged posteriorly, with 7 bristles (5 anterior separated by space from 2 stout plumose bristles).

Seventh Limb (Figure 28*i*): Each limb with 2 proximal bristles, 1 on each side with 1 or 2 bells, and 6 terminal bristles with 2 to 4 bells; bristles slightly tapered. Terminus with comb of about 6 curved teeth opposite 2 small pegs.

Furca: Similar to that of adult female.

Bellonci Organ: Broken; elongate and annulate (Figure 28*c,k,l*).

Eyes: Lateral eye with 3 or 4 ommatidia (Figure 28*j, l*). Medial eye larger than lateral eye, with some brownish pigmentation (Figure 28*k,l*).

Upper Lip: Not observed.

Posterior of Body: Bare.

Y-Sclerite: Similar to that of adult female.

Genitalia: None observed.

Eggs: BPBM-S 12849with 1 round unextruded egg (diameter 0.08 mm).

Description of instar IV male (Figure 29). Carapace shape and ornamentation similar to that of adult female (Figure 29*a*).

Infold: Rostral infold with row of 6 bristles and a pair of bristles at inner edge of incisure. Anteroventral infold with 1 bristle near inner end of incisure and about 7 narrow ridges. Infold of caudal process with 7 frondlike bristles along ridge forming pocket. Minute bristle at posterior tip of caudal process. Additional bristles along inner edge of posteroventral infold anterior to caudal process.

Central Adductor Muscle Scars (Figure 29*b,c*): Numerous elongate and ovoid muscle scars.

Selvage: Similar to that of adult female.

Carapace Size (length, height in mm): USNM 1085956, 0. 83, 0.45.

First Antenna: Similar to that of adult female.

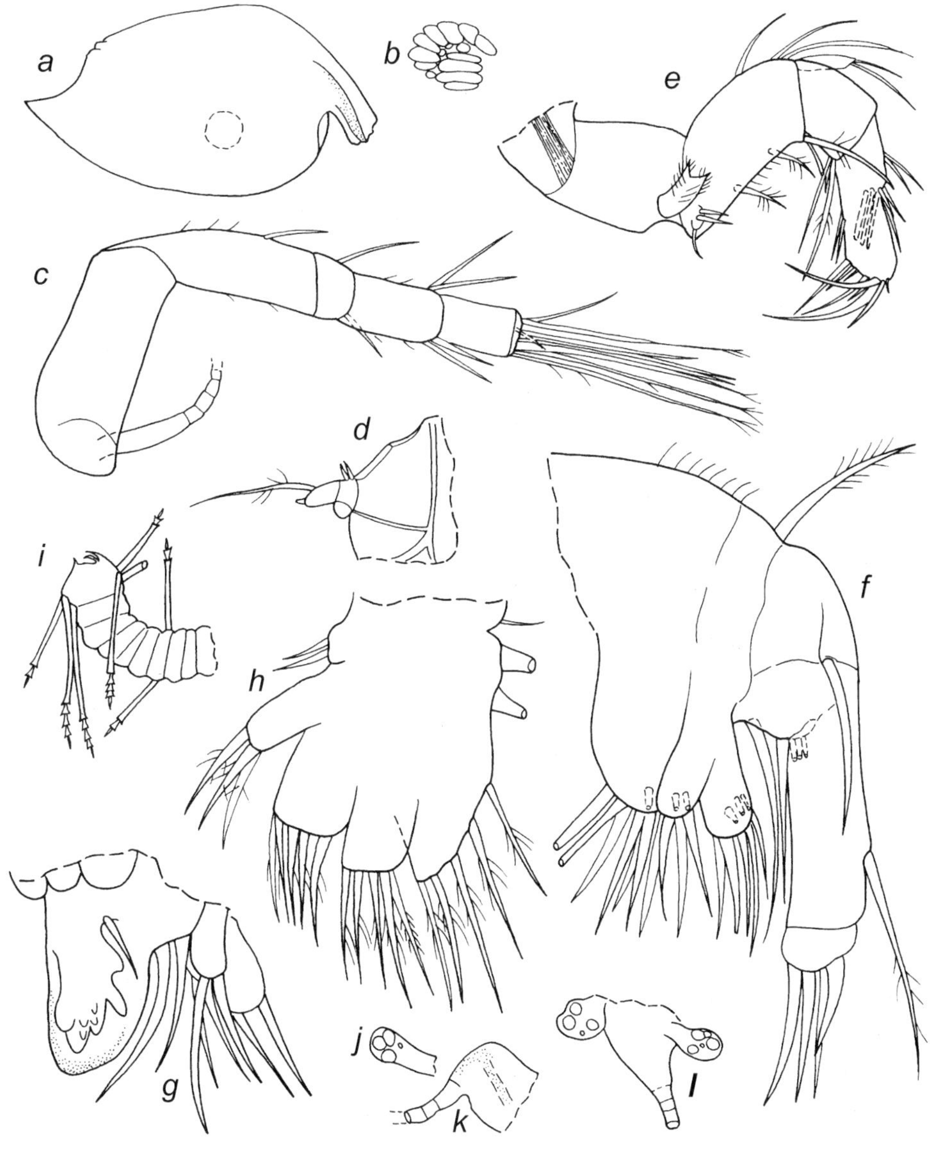

Figure 28. *Harbansus hox*, new species, paratype, BPBM-S 12849, female instar IV: *a,* complete specimen from right side, length 0.84 mm; *b,* central adductor muscle attachments right valve, ov; *c,* proximal part of Bellonci Organ and right 1st antenna, lv; *d,* endopod right 2nd antenna, mv; *e,* left mandible, mv; *f,* left maxilla, mv (nabs); *g,* distal part of 5th limb; *h,* 6th limb; *i,* 7th limb; *j,* left lateral eye, *k,* medial eye and proximal part of Bellonci Organ; *l,* dorsal view of lateral eyes, medial eye, and proximal part of Bellonci Organ.

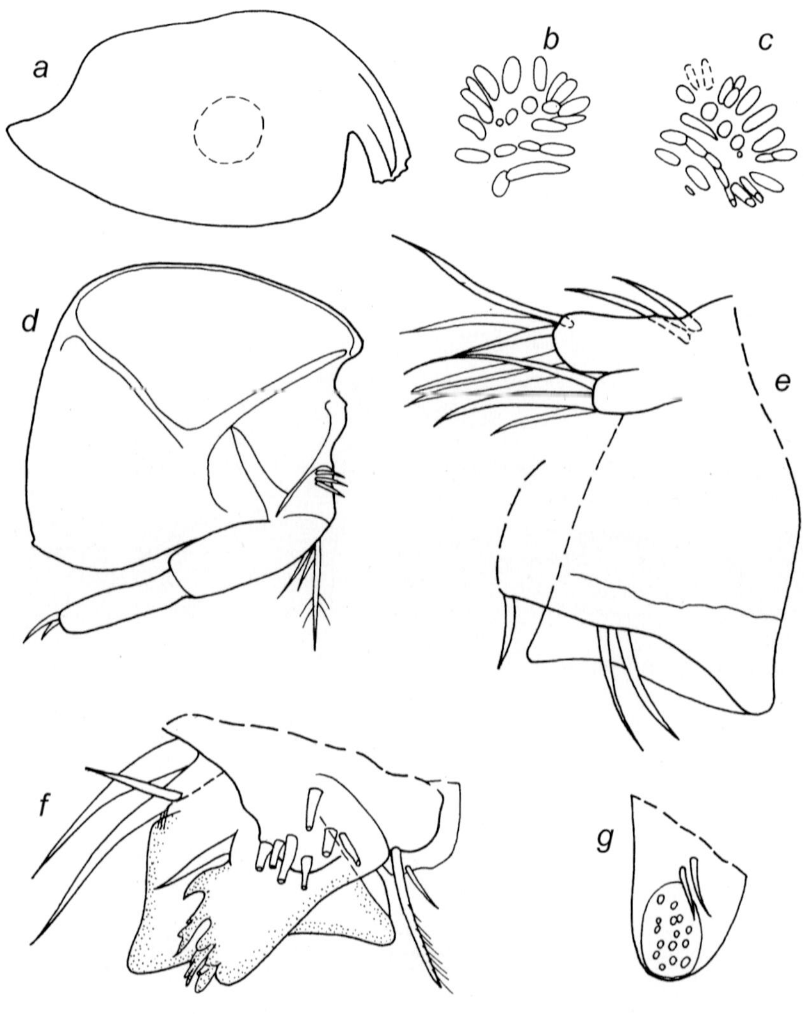

Figure 29. *Harbansus hox*, new species, paratype, USNM 1085956, male instar IV: *a,* complete specimen from right side, length 0.83 mm; *b,* central adductor muscle attachments of right valve, ov; *c,* central adductor muscles protruding from left side of body, anterior to left; *d,* protopod and endopod left 2nd antenna, mv; *e, f,* distal right and left 5th limbs (nabs), av; *g,* right copulatory organ.

Second Antenna: Protopod and exopod similar to that of adult female (exopodial bristles long with distal natatory hairs). Endopod with 3 articles (Figure 29*d)*: article 1 with 2 bristles; article 2 elongate with 3 bristles (1 long spinous, 2 short bare); article 3 elongate with 2 short terminal bristles.

Mandible: Coxa endite bifurcate, spinous. Basis: Medial side spinous and with 3 small bristles in proximal ventral corner, 1 short bristle near midlength with base set back from ventral margin; ventral margin with 2 bristles near midlength and 1 long spinous subterminal bristle; dorsal margin with 1 bristle near midlength and 2 terminal bristles. Exopod hirsute, about one-half length of 1st endopod article, with long subterminal bristle and short terminal bristle. Endopod: 1st article with 2 long and 2 short terminal ventral bristles. 2nd article: ventral margin with bristles forming 2 distal groups (proximal group with 2 bristles, distal group with 3 bristles including 1 clawlike bristle); dorsal margin with bristles forming 2 groups near midlength (2 bristles in proximal group, 6 in distal group). 3rd article with 2 stout claws and 3 bristles (no minute dorsal claw observed).

Maxilla Not examined in detail, but, in general, similar type to that of adult female.

Fifth Limb (Figure 29*e,f*): In general, similar to that of adult female.

Sixth Limb: Same as adult female: With 1 epipod bristle. Endite I with 3 bristles; endite II with 3 bristles; endite III with 6 bristles; endite IV with 5 bristles. End article not prolonged posteriorly, with 7 bristles (5 anterior spinous bristles separated by space from 2 stout plumose bristles).

Seventh Limb: Each limb with 4 terminal bristles with 4 bells; bristles tapered. Terminus with comb of about 5 curved teeth opposite 2 small pegs.

Furca: Similar to that of adult female.

Bellonci Organ: Not observed.

Eyes: Lateral eye small with 4 or 5 ommatidia. Medial eye larger than lateral eye, with some brownish pigmentation.

Upper Lip: Not observed.

Posterior of Body: Bare.

Y-Sclerite: Similar to that of adult female.

Genitalia (Figure 29*g*): Lobe on each side of body with 2 small proximal bristles. Lobe located just posterior to base of 6th limb.

Comparisons. The new species *H. hox* differs from *H. barnardi* and *H. vix* in having long natatory bristles on the exopod of the 2nd antenna of the adult female.

Superfamily CYLINDROLEBERIDOIDEA Müller

The Cylindroleberidoidea contains the family Cylindroleberididae Müller.

Family CYLINDROLEBERIDIDAE Müller

This family includes 3 subfamilies, Cylindroleberidinae Müller, Cyclasteropinae Poulsen, and Asteropteroninae Kornicker, of which only the Cylindroleberidinae are represented in the present collection. Kornicker (1976: 15) described *Microasteropteron youngi* Kornicker from Maunalua Bay, O'ahu. That species is in the Cyclasteropinae.

Subfamily CYLINDROLEBERIDINAE Müller

This subfamily includes 2 tribes, Cylindroleberidini Müller and Bruuniellini Kornicker & Harrison-Nelson, of which only the former is represented in the present collections.

Tribe CYLINDROLEBERIDINI Müller

This tribe contains 18 genera, of which 3 are represented in the present collections.

Genus *Parasterope* Kornicker

Parasterope Poulsen, 1965: 361. *Nomen nudum*; genus-group name proposed after 1930 without type species designation.

Parasterope Kornicker, 1975: 401. **Type species**. *Asterope muelleri* Skogsberg, 1920, by original designation.

Composition and distribution. This genus contains numerous species. Widespread between latitudes of about 55°N and 65°S. Depth range intertidal to 4303 m (Kornicker & Caraion, 1974: 7).

Remarks concerning non-types. Six new species of *Parasterope* are described herein. The species have similar carapaces in shape and size, but could be identified with certainty by small differences in appendage morphology. In order to have whole specimens for possible future study by others, a few have been left unopened. These have been referred to identified species in the same sample and are designated "non-types". Specimens from French Frigate Shoals have also been designated non-types.

KEY TO SPECIES OF *PARASTEROPE* FROM O'AHU AND FRENCH FRIGATE SHOALS

1. Mandible with small bristle at midheight of dorsal margin of basis .... ***P. omega*, n. sp.**
–. Mandible without small bristle at midheight of dorsal margin of basis 2

2. Mandible with small bristle (non-triaenid type) proximal to U-shaped process on ventral margin of basis 3
–. Mandible without small bristle proximal to U-shaped process on ventral margin of basis ... 4

3. Exopod of 2nd antenna with basal spines .... ***P. zeta*, n. sp.**
–. Exopod of 2nd antenna without basal spines .... ***P. theta*, n. sp.**

4. Exopod of 2nd antenna with basal spines 5
–. Exopod of 2nd antenna without basal spines .... ***P. gamma*, n. sp.**

5. Second antenna with rows of small medial spines in proximal dorsal corner of protopod .... ***P. sigma*, n. sp.**
–. Second antenna without rows of small medial spines in proximal dorsal corner of protopod ... . ***P. iota*, n. sp.**

***Parasterope gamma*, new species**

Figs. 30–32

Holotype. BPBM-S 12861, ovigerous female on slide and in alcohol.

Type Locality. Sta 10, 22 Jan 2001, Kapua Channel, Waikīkī, O'ahu, Hawaiian Islands.

Paratypes. None.

Non-types. USNM 1085957, ovigerous female on slide and in alcohol, French Frigate Shoals, Sta TC-12.

Distribution. Waikīkī, Sta 10. French Frigate Shoals, Sta TC-12.

Description of adult female (Figures 30–32). Carapace in lateral view wider in posterior part (Figures 30*a,* 32*d*).

Infold: Rostral infold plus anterodorsal infold with 21 to 27 bristles (Figure 30*c*). Anteroventral infold with 19 long bristles and 3 or 4 minute bristles (Figure 30*b*). Small bristle at inner end of incisure (Figure 30*b*). Ventral margin with row of 8 to 13 bristles. Posterior infold with broad band with 19 broad flaplike bristles and 7 or 8 short slender bristles (not more than 1 between some pairs of flaplike bristles) (Figure 30*d*). Anteroventral part of posterior infold with 4 to 7 long bristles between broad band and posteroventral valve edge (Figure 30*d*). Valve edge with minute pores. Posterior infold of valves without narrow ridge between ridge bearing flaplike bristles and posterior margin of valve.

Central Adductor Muscle Attachments (Figure 30*e*): Each valve with about 10 oval attachments.

Selvage: None observed in vicinity of incisure or along free margins.

Carapace Size (length, height in mm): BPBM-S 12861 , 0.91, 0.58. USNM 1085957, 0.93, 0.61.

Figure 30. *Parasterope gamma,* new species, holotype, BPBM-S 12861, adult female: *a,* complete specimen, length 0.91 mm; *b, c,* anterior left valve, iv; *d,* posterior left valve, iv; *e,* some central adductor muscle attachments right valve, ov; *f,* right 1st antenna, lv; *g,* protopod and endopod left 2nd antenna, mv; *h,* tip exopod right 2nd antenna, lv; *i,* coxa endite left mandible, mv; *j,* comb bristles 5th limb (nabs).

First Antenna (Figure 30*f*): 1st article bare. 2nd article with few dorsal spines, long distal spinous dorsal bristle, and short distal lateral bristle. 3rd and 4th articles separated by suture better developed on medial side. 3rd article: short ventral margin with small bristle; long dorsal margin with few minute spines and 6 bristles (2 with bases lateral, 4 with bases along dorsal edge; 4th bristle along edge with short spines, other 3 bristles along edge with long proximal spines; 2 lateral bristles with few long spines). 4th and 5th articles separated by concave suture. 4th article with 3 terminal bris-

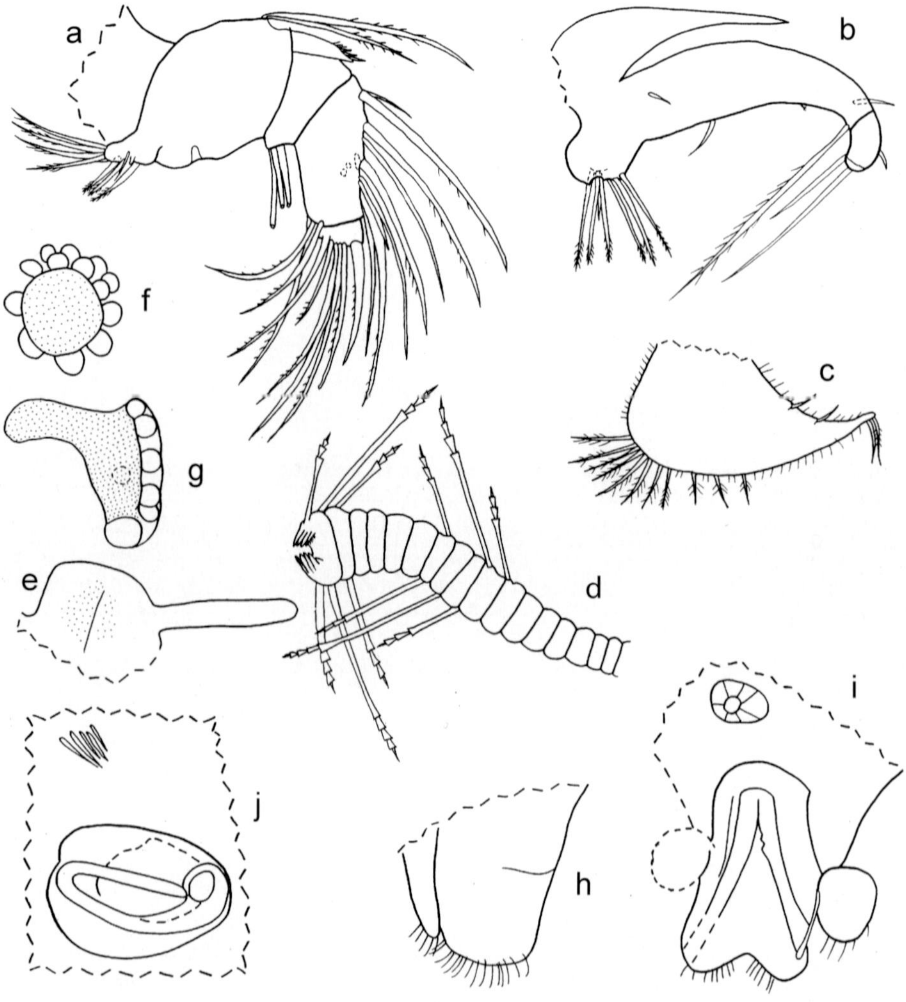

Figure 31. *Parasterope gamma,* new species, holotype, BPBM-S 12861, adult female: *a,* right mandible, lv; *b,* right maxilla, lv; *c,* right 6th limb, lv; *d,* 7th limb; *e,* medial eye and Belloncì Organ from right side; *f,* left lateral eye seen through valve; *g,* right lateral eye, pv; *h,* upper and lower lips from right side; *i,* upper and lower lips and mouth, pv; *j,* genitalia and adjacent bristle, viewed from right side.

tles (2 ventral, 1 dorsal). Sensory bristle of 5th article with 6 filaments with all bases close to tip. 6th article with fairly long medial bristle. 7th article: a-bristle clawlike; b-bristle with 3 marginal filaments; c-bristle long with 4 marginal filaments. 8th article: d-bristle absent; e-bristle longer than a-bristle, bare with blunt tip; f-bristle bent dorsally; g-bristle long, with 4 marginal filaments. Total bristles 20.

Second Antenna (Figure 30*g,h*): Protopod with small distal medial bristle (Figure 30*g*). Endopod with 3 articles: 1st and 2nd article indicated by indentation in edges, but without suture; 2nd and 3rd articles separated by indistinct, but visible, suture; 3rd article with long terminal filament (Figure 30*g*). Exopod with 9 articles (Figure 30 *h*): bristle of 2nd article reaching past 9th article, with ventral spines; bristles of articles 3 to 8 with long bristles with proximal minute ventral spines and distal natatory hairs; 9th article with 3 bristles (dorsal bristle short bare, middle bristle with indistinct ventral spines and distal indistinct hairs, longer ventral bristle with natatory hairs). Exopod articles without basal spines. Article 9 with lateral spine with digitate tip. Articles 2 to 8 with small terminal spines on terminal dorsal corner. Total bristles 13.

Mandible (Figures 30*i*, 31*a*): Coxa endite (Figure 30*i*): small bristle at base of ventral branch; ventral branch with 3 rows of spines and pointed tip with 3 minute spines; dorsal branch with 4 low nodes proximal to small main spine; tip with small terminal spine and short dorsal bristle. Basis endite with 4 spinous end bristles, 3 triaenid bristles with 3 to 5 pairs of spines proximal to terminal pair, fairly long dwarf bristle and glandular peg. Ventral margin of basis with U-shaped process, no bristles; dorsal margin with 2 long terminal bristles. Exopod almost as long as dorsal margin of 1st article of endopod, with dorsal hairs and 2 short subterminal bristles. Endopod: 1st article with 3 long spinous ventral bristles (shortest with short marginal spines, others with long spines near midlength). 2nd article: dorsal margin with 1 short slender proximal bristle followed by long spinous a-, b-, c-, and d-bristles (c-bristle slightly stouter than others); lateral side of dorsal margin with 1 long bristle between b- and c-bristles, and 1 long bristle between c- and d-bristles; medial side near dorsal margin with 4 to 5 short cleaning bristles (0 or 1 between a- and b-bristles and 4 forming row between c- and d-bristles), and 1 long bristle adjacent to d-bristle; ventral margin with 3 long bristles with short marginal spines (medial of the bristles shorter than others). 3rd endopod article with stout terminal claw with minute ventral spines, 3 long terminal bristles with short marginal spines, 1 short medial terminal spinous bristle near ventral margin, and 1 long spinous medial bristle just ventral to base of dorsal claw. Total bristles 36 or 38.

Maxilla (Figure 31*b*): Endite I with 4 bristles (3 long, 1 short), endite II with 3 long bristles. Epipod with pointed tip reaching about midlength of dorsal margin of basis. Basis with short proximal lateral bristle near midheight, distal short medial bristle near dorsal margin, proximal backward pointing short ventral bristle, minute distal ventral bristle, and long spinous terminal ventral bristle. Ventral margin of comb with 71 bristles in addition to 1 proximal bristle with bent tip. Endopod: 1st article with short dorsal bristle and long terminal beta-bristle; 2nd article with terminal bristle longer than beta-bristle. Total bristles approximately 86.

Fifth Limb (Figure 30*j*): Epipod with 50 bristles. Comb with long exopod bristle, 1 short slender bristle ventral to base of exopod bristle, and 3 short bristles close to ventral margin (margin obscure).Ventral margin of comb with about 29 bristles. Oblique section anterior to epipod with about 21 bristles. Total bristles approximately 105.

Sixth Limb (Figure 31*c*): Anterior margin with 2 endite bristles. Ventral margin with 2 spinous anterior bristles separated by space from 11 bristles (anterior 3 bristles shorter and with long proximal and short distal spines, others with long spines). Anterior, ventral, and posterior margins, and medial surfaces spinous. Lateral flap spinous but without bristles. Minute medial bristle in anterodorsal corner not observed with certainty. Total bristles 15.

Seventh Limb (Figure 31*d*): With 6 proximal bristles (3 on each side) and 6 terminal bristles (3 on each side); each bristle with 3 or 4 bells. Terminus with opposing combs, each with 8 or 9 spinous teeth. Total bristles 12.

Furca (Figure 32*a,c*): Each lamella with 7 claws (last claw could be interpreted to be a bristle) and 2 backward pointing bristles. Claws with small teeth along posterior edges; claws 1–3 with hairs along anterior edges. Right lamella anterior to left by width of base of claw 1. Total bristles and claws 9.

Bellonci Organ (Figure 31*e*): Elongate with rounded tip.

Eyes: Lateral eye with about 17 ommatidia and black pigment (Figures 31*f, g*, 32*e*). Medial eye about same size as lateral eye, with light amber color.

Gills (Figure 32*b,c*): Posterior end of gills not reaching past posterior end of body; exact number not determined. Gills of BPBM-S 12861 unusual in having 3 or 4 nodes along posterior edges; nodes on posterior gill appears somewhat sclerotized and nodes more strongly developed than on other gills.

Lips (Figure 31*h,i*): Upper lip rounded, hirsute, and with small anterior spine. Lower lip rounded, hirsute. Mouth round in posterior view (Figure 31*i*).

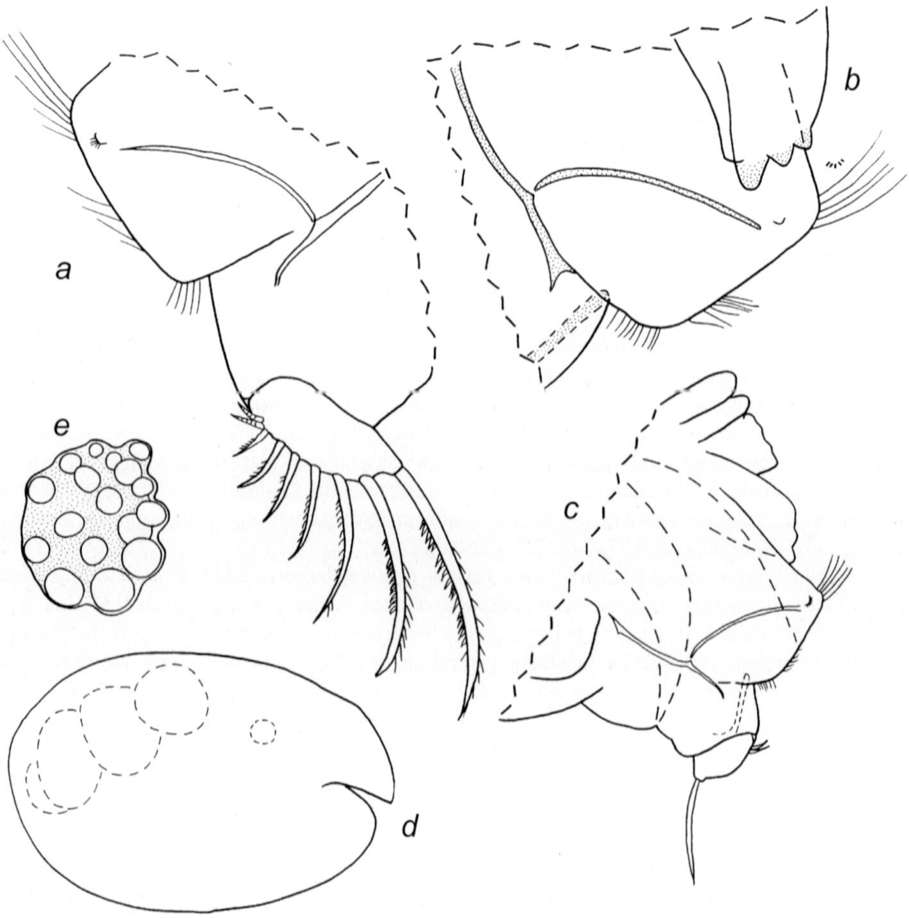

Figure 32. *Parasterope gamma,* new species, holotype, BPBM-S 12861, adult female: *a,* posterior of body from right side; *b, c,* posterior of body from left side. USNM 1085956, adult female, non-type: *d*, complete specimen, length 0.93 mm; *e*, lateral eye.

Posterior of Body (Figure 32*a–c*): Posterodorsal corner with very long hairs; shorter hairs along margin ventral to corner. A single spinous crescent present near posterodorsal corner in vicinity of posterior end of girdle.

Y-Sclerite (Figure 32*a–c*): Elongate with pointed tip.

Genitalia (Figure 31*j*): Oval with coiled structures on each side of body adjacent to 5 indistinct bristles and anterior to furca.

Eggs in Marsupium (Figures 30*a,* 32*d*): BPBM-S 12861 with 4 eggs in brood chamber, and may have few very small unextruded eggs, but could not identify them with certainty. USNM 1085957 with 5 eggs in brood chamber.

Comparisons (adult female). The exopod of the 2nd antenna of *P. gamma* differs from that of *P. pacifica* Kornicker & Harrison-Nelson in lacking basal spines on articles 4–8. The posterodorsal part of the body of the *P. gamma* just dorsal to the posterior end of the girdle bears 1 minute spinous crescent compared to 4 on *P. pacifica. P. muelleri* (Skogsberg) has 3 crescents on the posterodorsal part of the body, and also basal spines on articles 4–8 of the exopod of the 2nd antenna. The dorsal margin of the basis of the mandible of *P. gamma* is without the midbristle present on *P. zamboangae* Kornicker.

Parasterope omega, new species

Figs. 33–36

Holotype. BPBM-S 12841, ovigerous female on 2 slides and in alcohol.

Type locality. Sta 6, 19 Nov 1999, Pu'u Kaua'i reef, Kāne'ohe Bay, O'ahu, Hawaiian Islands.

Paratypes. Kāne'ohe Bay: Sta 1, North Channel: BPBM-S 12838, ovigerous female with body removed from shell in alcohol. Waikīkī: Sta 12, Ag. reef, USNM 1085958, partly dissected ovigerous female on 2 slides and in alcohol.

Non-types. Kāne'ohe Bay: Sta 6 (type locality), USNM 1085959, undissected adult female in alcohol. Sta 1, North Channel, BPBM-S 12839, undissected ovigerous female in alcohol.

Distribution. Kāne'ohe Bay: Sta 1,6. Waikīkī: Sta 12.

Description of adult female (Figures 33–36). Carapace elliptical or slightly tumid in lateral view (Figures 33*a*, 36*a*,*f*).

Infold: Rostral infold with 42 bristles (Figure 33*d*); 1 small bristle present at inner end of incisure; anteroventral infold with 28 bristles (Figure 33*e*). Ventral infold to point just anterior to posterior ridge bearing triangular flaplike bristles with 13 bristles. Posteroventral infold distal to ridge bearing flaplike bristles with 8 bristles between ridge and posterior valve margin; 1 small bristle present opposite dorsal end of ridge bearing flaplike bristles (Figure 33*f*). Posteroventral and posterior ridge with 27 flap like bristles and about 17 minute bristles (1 or 2 bristles between some flaplike bristles). Dorsal end of ridge bearing flaplike bristles terminates at point just dorsal to midheight of dorsal margin of valve. Left valve of holotype with narrow ridge bearing pores along posterior valve edge. Right valve of holotype with narrow ridge bearing minute pores or bristles between ridge bearing flaplike bristles and posterior margin of valve (Figure 33*f*). Pore-bearing ridge on both valves of paratype USNM 1085958 in similar location as that of left valve of holotype.

Selvage: None observed in vicinity of incisure or along free margins.

Central Adductor Muscle Attachments (Figure 33*a*–*c*): Consisting of 12 ovoid attachments.

Carapace Size (length, height in mm): Sta 6: BPBM-S 12841, 1.10, 0.71; USNM 1085959, 1.09, 0.71. Sta 1: BPBM-S 12838, 1.08, 0.71; BPBM-S 12839, 1.04, 0.73. St. 12: USNM 1085958, 1.08, 0.69.

First Antenna (Figure 34*a*): 1st article with spines forming row in distal ventral half. 2nd article with ventral spines, few dorsal hairs, 1 spinous dorsal bristle, and 1 distal lateral bristle at midheight. 3rd article with short ventral margin with small bristle, and long dorsal margin with 6 spinous bristles (4 bristles with bases on edge with spines (proximal 3 with long spines, distal with short spines), 1 bristle (with long spines) with base lateral between 2nd and 3rd edge bristles, and 1 bristle (with long spines) lateral to distal edge bristle). 3rd and 4th articles separated by oblique slightly curved suture. 4th article: long ventral margin with few spines and 2 short terminal bristles, short dorsal margin with long bristle with long spines. Long 5th article with minute dorsal spines and terminal ventral sensory bristle with long proximal part and 6 long terminal filaments. Long 6th article with long medial bristle. Short 7th article with clawlike a-bristle with minute proximal dorsal spines; b-bristle longer than a-bristle, with 3 long marginal filaments; c-bristle long with 4 marginal filaments. 8th article fused to 7th article: d-bristle absent; e-bristle longer than a-bristle, bare with blunt tip; f-bristle bent dorsally, with 4 marginal filaments (3 proximal, 1 distal); g-bristle long with 5 marginal filaments. Total bristles 20.

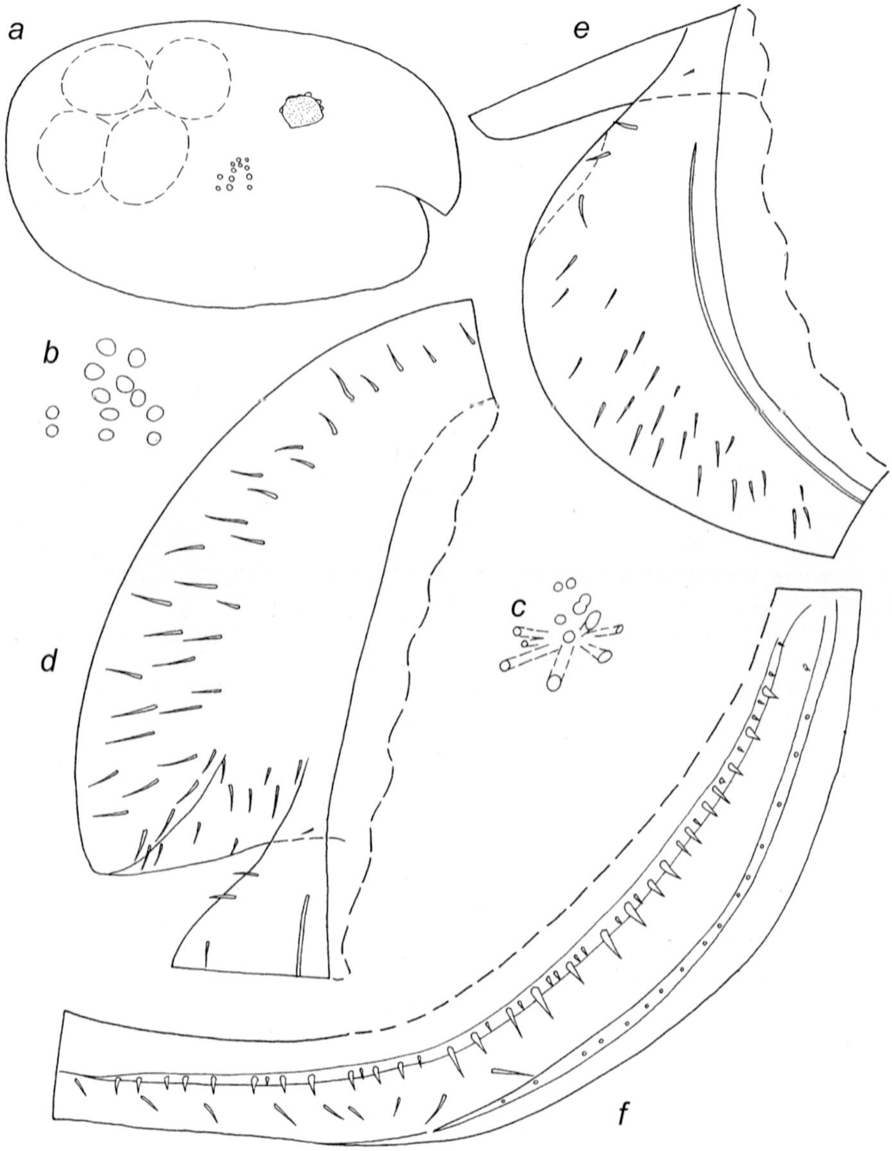

Figure 33. *Parasterope omega,* new species, holotype, BPBM-S 12841, adult female: *a,* complete specimen from right side, length 1.10 mm; *b,* detail of central adductor muscle attachments, from "*a*"; *c*, central adductor muscles projecting from right side of body with right valve removed; *d*, *e*, anterior right valve, iv; *f*, posterior right valve, iv.

Second Antenna: Protopodite with small distal medial bristle and few spines along distal dorsal margin (Figure 34*b*). Endopod with 3 articles (1st and 2nd articles fused); 3rd article with long filament (Figure 34*b*). Exopod: bristle of 2nd article reaching just past 9th article, with row of long slender ventral spines; bristles of articles 2 to 8 long with proximal ventral spines and natatory hairs; 9th article with 4 bristles: 1 long (ventral), with proximal ventral spines and natatory hairs; 1 medium with natatory hairs, 2 short (dorsal) bare. Article 1 with long stout spines forming distal row (Figure 34*b*).

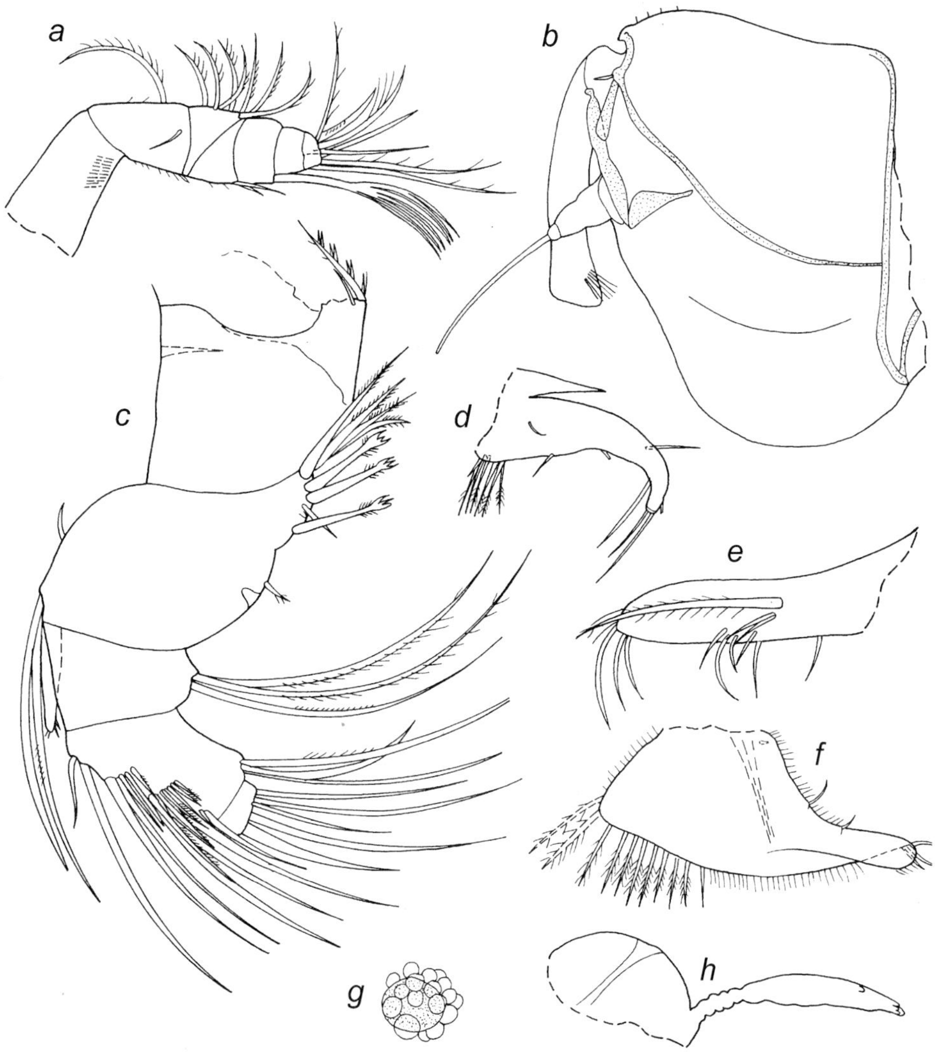

Figure 34. *Parasterope omega,* new species, holotype, BPBM-S 12841, adult female: *a,* right 1st antenna, lv; *b,* protopod and endopod right 2nd antenna, mv; *c*, right mandible, mv; *d*, right maxilla, lv (nabs); *e*, comb left 5th limb, lv; *f*, right 6th limb, lv; *g*, right lateral eye; *h*, medial eye and Bellonci Organ.

Articles 3 to 8 with indistinct minute spines at terminal dorsal corners. Articles 5 to 8 with short basal spines (spine of article 8 about length of 9th article); Article 9 with short lateral spine, about one-half length 9th article. Sclerites in vicinity of endopod of usual type (Figure 34*b*). Total bristles 13.

Mandible: (Figures 34*c*, 35*c*, 36*b,c*): Coxa endite of holotype broken off and lodged inside of esophagus (Figure 35*c*), with small medial bristle near base of ventral branch (Figure 34*c*). Coxa endite of paratype USNM 1085958: ventral branch with 4 rows of ventral spines and elongate tip with 2 or 3 minute indistinct spines at tip; dorsal branch with long subterminal dorsal bristle and small terminal bristle; small medial bristle present near base of ventral branch (Figure 36*b*). Basis endite with 3 triaenid bristles with 3 or 4 pairs of distal spines plus terminal pair, 1 dwarf bristle adjacent to peg, and 4 spinous end bristles. Basis: ventral margin with 1 small spinous bristle proximal

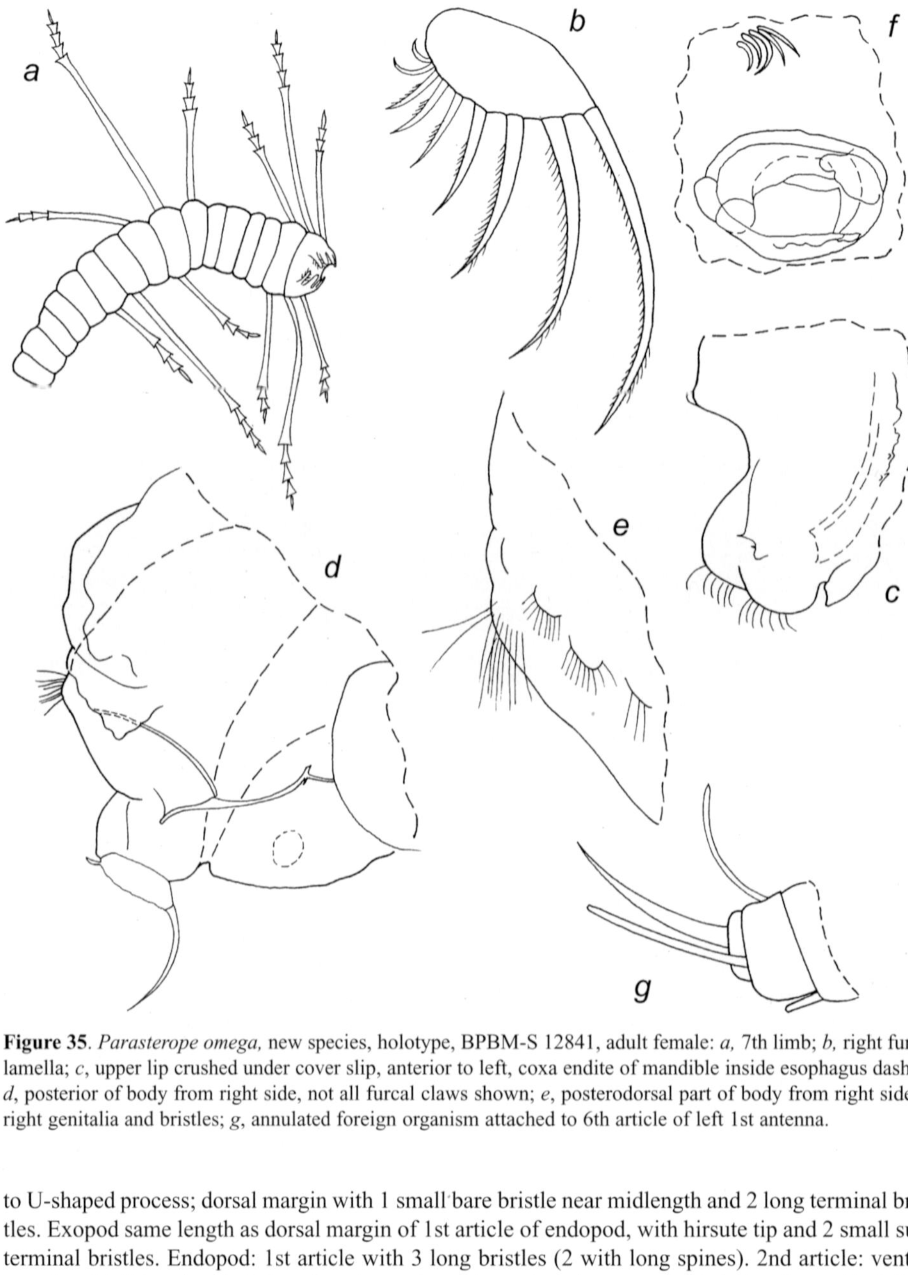

Figure 35. *Parasterope omega,* new species, holotype, BPBM-S 12841, adult female: *a,* 7th limb; *b,* right furcal lamella; *c*, upper lip crushed under cover slip, anterior to left, coxa endite of mandible inside esophagus dashed; *d*, posterior of body from right side, not all furcal claws shown; *e*, posterodorsal part of body from right side; *f*, right genitalia and bristles; *g*, annulated foreign organism attached to 6th article of left 1st antenna.

to U-shaped process; dorsal margin with 1 small bare bristle near midlength and 2 long terminal bristles. Exopod same length as dorsal margin of 1st article of endopod, with hirsute tip and 2 small subterminal bristles. Endopod: 1st article with 3 long bristles (2 with long spines). 2nd article: ventral margin with 3 terminal bristles; dorsal margin with stout a-, b-, c-, and d-bristles (base of c-bristle about same width as base of d-bristle) and 1 small proximal bristle; lateral surface with 1 long bristle between b- and c-bristles and between d- and e-bristles; medial surface with 3 cleaning bristles forming oblique row between b- and c-bristles, 5 cleaning bristles forming oblique row between c- and d-bristles, and 1 fairly long bristle adjacent to d-bristle. 3rd article with stout dorsal clawlike bristle, 3 long end bristles, 1 short medial bristle near ventral margin, and 1 long lateral bristle just ventral to dorsal clawlike bristle. Total bristles 43 (does not include 2 bristles present on coxa endite).

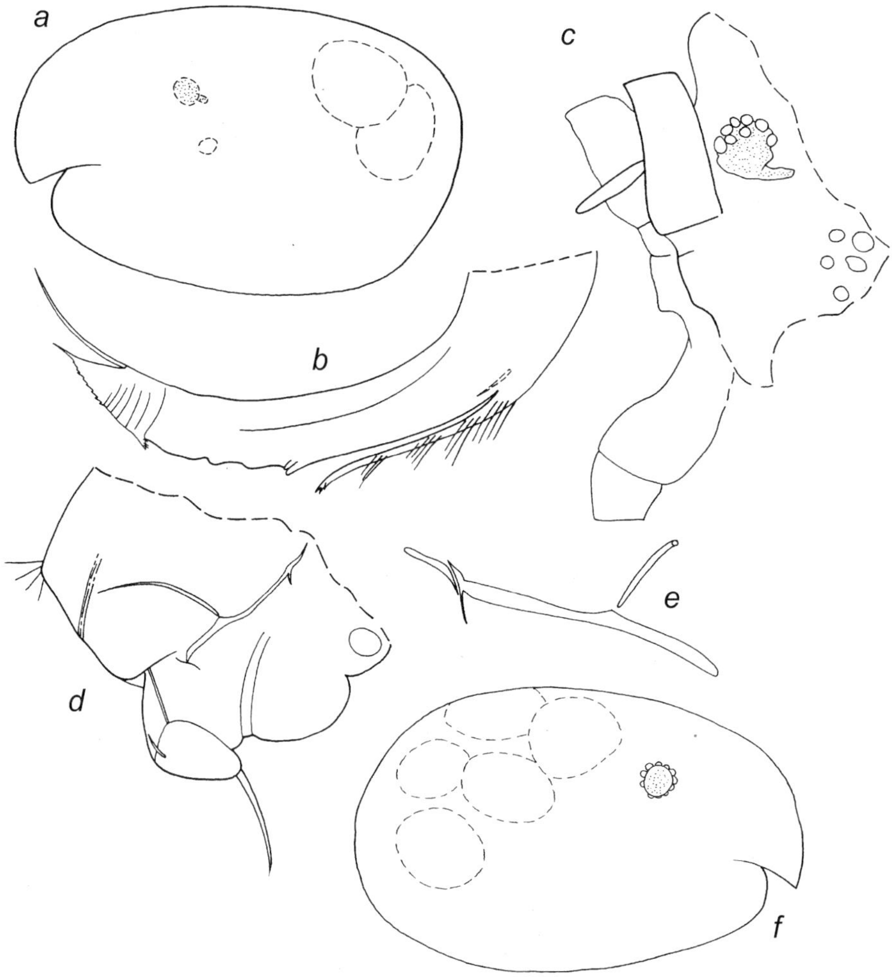

Figure 36. *Parasterope omega,* new species, paratype, USNM 1085958, adult female: *a,* complete specimen from left side, length 1.08 mm; *b,* coxa endite right mandible, lv; *c*, anterior of body from left side; *d*, posterior of body from right side, not all furcal claws shown; *e*, left Y-Sclerite and ventral end of girdle, anterior to left. Paratype, BPBM-S 12838, adult female: *f*, complete specimen from right side, length 1.08 mm.

Maxilla (Figure 34*d*): Epipod reaching to about midlength of basis. Endite I with 1 short bristle and 3 long spinous bristles. Endite II with 2 or 3 long bristles. Basis: lateral surface with short proximal bristle; ventral margin with 1 proximal backward pointing bristle, 1 small distal bristle, and 1 long spinous end bristle; dorsal margin with fairly long distal bristle. Endopod: 1st article with small alpha-bristle and long beta-bristle; 2nd article with terminal bristle reaching past beta-bristle. Comb with 75 bristles excluding proximal bent bristle. Total bristles 87.

Fifth Limb (Figure 34*e*): Comb with spinous exopod bristle reaching past tip of comb, 2 slender bristles just ventral to base of exopod bristle, 4 bristles near midlength of comb and close to ventral edge, 1 proximal bristle and 2 distal bristles with bases almost on comb edge. Epipod with 55 bristles, oblique section with about 22 bristles, ventral edge of comb with 27 bristles. Approximate total bristles 114.

Sixth Limb (Figure 34*f*): Anterior margin with upper and lower endite bristles. Proximal anterior corner with 1 minute medial bristle. Anterior corner of ventral margin with 2 bristles. Lateral flap with long hairs but no bristles. Posterior half of ventral margin with 10 or 11 spinous bristles; bristles along posteroventral curvature longer. Anterior, ventral, and posterior margins hirsute. Total bristles 15 or 16.

Seventh Limb (Figure 35*a*): Proximal group with 6 bristles, 3 on each side, each with 3 or 4 bells; terminal group with 6 bristles, 3 on each side, each with 2 to 4 bells. Terminous with opposing combs, each with about 6 spinous teeth. Total bristles 12.

Furca (Figures 35*b, d,* 36*d*): Each lamella with 6 claws followed by 3 bristlelike claws. Total claws and bristles 9.

Bellonci Organ: Elongate with crinkled proximal one-fourth and small subterminal protuberance on holotype (Figure 34*h*). Subterminal protuberance absent on paratype USNM 1085958 (Figure 36*c*).

Eyes: Lateral eye with 17 ommatidia mostly obscured by black pigment (Figures 34*g,* 36*a,c,f*). Medial eye bare without pigment, about same size as lateral eye (Figure 34*h*).

Gills (Figure 35*d*): Well developed, some with uneven posterior edge.

Lips: Hirsute lobes (Figure 35*c* shows coxa endite of mandible inside esophagus).

Anterior of Body (Figures 35*c,* 36*c*): With small anterior protuberance on each side between 1st antennae and mandibles.

Posterior of Body (Figures 35*d,e,* 36*d*): Spinous with rounded posterodorsal corner and 3 spinous crescentlike processes anterior to corner (Figure 35*e*).

Y-Sclerite (Figures 35*d,* 36*d,e*): Elongate with slight expansion proximal to pointed tip and ventral spinelike process proximal to tip.

Genitalia (Figure 35*d,f*): Oval area containing internal coiled elongate structures on each side of body anterior to furca; 5 minute bristles present dorsal to genital area

Eggs (Figures 33*a,* 36*a, f*): BPBM-S 12841 with 5 eggs in marsupium, length of 1 egg 0.25 mm. BPBM-S 12838 with 8 eggs in marsupium and unextruded small eggs, length of 1 egg 0.26 mm. USNM 1085958 with 2 eggs in marsupium.

Gut Content: Minute amber colored particles.

Foreign Organisms: BPBM-S 12841 with annulated filaments attached to 1st antenna (Figure 35*g*), mandible, and 6th limb.

Comparisons. The dorsal margin of the basis of the mandible of *P. omega* differs from those of *P. gamma* and *P. pacifica* in having a midbristle. The ventral margin of the basis of the mandible of *P. omega* differs from that of *P. zamboangae* in having a plain bristle and not a triaenid bristle proximal to the U-shaped process. The comb of the 5th limb of *P. omega* differs from that of *P. gamma* in having 2 rather than 1 slender bristle just ventral to the base of the stout exopod bristle.

Parasterope sigma, new species

Figs. 37–39

Holotype. BPBM-S 12840, adult female on 2 slides and in alcohol.

Type-locality. Sta 1, 17 Nov 1999, North Channel, Kāne‘ohe Bay.

Paratypes. None.

Distribution. North Channel, Kāne‘ohe Bay.

Description of adult female (Figures 37–39). Carapace elliptical in lateral view (Figure 37*a*). Vestment with abundant minute foreign spheres.

Infold: Rostral infold with about 28 bristles (Figure 37*c*); broad part of anteroventral infold with about 20 bristles; minute bristle at inner margin of incisure; ventral infold with about 20 bristles. Infold of caudal process with inner ridge bearing 28 to 33 flaplike bristles (Figure 37*d*); about 8 minute bristles present between flaplike bristles (not more than 2 bristles between a pair of flaplike bristles). About 7 bristles between the ridge bearing flaplike bristles and ventral valve edge.

Selvage: None observed in vicinity of incisure or along free margins.

Central Adductor Muscle Attachments: About 10 ovoid scars (not all shown in Figure 37*b*).

Carapace Size (length, height in mm): BPBM-S 12840, 1.23, 0.81.

First Antenna (Figure 37*e*): Article 1 with lateral spines. Article 2 with spinous dorsal bristle and few proximal dorsal spines. Oblique suture between articles 3 and 4 well developed; article 3 with short ventral margin with small bristle and long dorsal margin with 6 bristles (4 bristles with bases on edge (proximal 3 with long spines, 4th with short spines), and 2 with long bases slightly lateral (1 between 2nd and 3rd edge bristles, 1 adjacent to 4th edge bristle); both bristles with long marginal spines). Suture between 4th and 5th articles well developed, concave; 4th article with short dorsal margin with 1 long bristle with short spines, and long ventral margin with few spines at midlength and 2 short terminal bristles. Sensory bristle of 5th article long slender with 6 long terminal filaments. Article 6 well developed with long terminal medial bristle with short marginal spines. Article 7: a-bristle clawlike, bare; b-bristle reaching past a-bristle, with 4 marginal filaments; c-bristle long with about 4 short marginal filaments. Article 8: d-bristle absent; e-bristle longer than a-bristle, bare with blunt tip; f-bristle bent dorsally, with 4 short marginal filaments.

Second Antenna: Protopod with short distal medial bristle and rows of minute medial spines near posterior dorsal corner (Figure 37*f*). Endopod with 3 articles well defined by sutures; 3rd article with long terminal filament (Figure 37*f*). Exopod: bristle of article 2 reaching past 9th article, with ventral spines; bristles of articles 3 to 7 with proximal ventral spines and distal natatory hairs; 9th article with 4 bristles (1 short dorsal bare, 1 longer with short marginal spines, 2 long with proximal ventral spines and distal natatory hairs). Articles 4 to 8 with basal spines (spine of 8th article about one-half length of 9th article and with rounded tip) (Figure 37*g*); article 9 with lateral spine almost length of article, with rounded tip on left limb of holotype and with indentation at midwidth of tip on right limb (Figure 37*g*). Articles 2 to 8 with minute spines at inner terminal corner.

Mandible: (Figures 38*a,b*, 39*f*): Coxa endite partly obscured on holotype; small medial bristle near base of ventral branch (Figure 38*b*). Basis endite with 3 or 4 spinous end bristles, 3 triaenid bristles with 3 or 4 pairs of marginal spines in addition to terminal pair, and fairly long dwarf bristle adjacent to small node (Figure 38*a*). Basis: dorsal margin with 2 long terminal dorsal bristles; ventral margin with U-shaped indentation. Exopod almost same length as dorsal margin of 1st article of endopod, hirsute distally and with 2 small subterminal bristles. Endopod: 1st article with 3 bristles (2 long with long marginal spines, 1 shorter with short marginal spines). 2nd article: ventral margin with 3 terminal bristles; dorsal margin with stout a-, b-, c-, and d-bristles and 1 short proximal bristle; lateral side with long bristle between b- and c-bristles and c- and d-bristles; medial side with 1 short bristle adjacent to b-bristle, diagonal row of 5 cleaning bristles between b- and c-bristles, and 1 short bristle adjacent to d-bristle. 3rd article with stout dorsal clawlike bristle with marginal spines, 3 long end bristles, 1 short medial bristle near ventral margin, and 1 long lateral bristle just ventral to base of dorsal clawlike bristle.

Maxilla (Figure 38*c,d*): Endite I with 4 bistles (3 long and 1 short). Endite II with 3 long bristles. Basis: lateral side with proximal bristle near dorsal margin, 1 distal dorsal bristle, 1 backward pointing proximal ventral bristle, 1 short distal ventral bristle, and 1 long spinous bristle at distal end of comb. Endopod: 1st article with short alpha-bristle and long beta-bristle; 2nd article with very long terminal bristle. Epipod broken off both limbs of holotype during dissection and not observed.

Fifth Limb: Comb (Figure 38*e*): spinous exopod bristle reaching past distal end of comb; 1 slender bristle just proximal to base of comb; 5 bristles closer to ventral edge of comb.

Sixth Limb (Figure 39*a*): Anterior margin with 2 endite bristles; anterodorsal corner with small medial bristle set back from anterior edge. Lateral flap with long hairs but no bristles. Anteroventral corner with 2 spinous bristles (anterior bristle about twice length of other bristle). Ventral margin with 18 spinous bristles; bristle longer along posteroventral curvature. Anterior, ventral, and posterior margins hirsute.

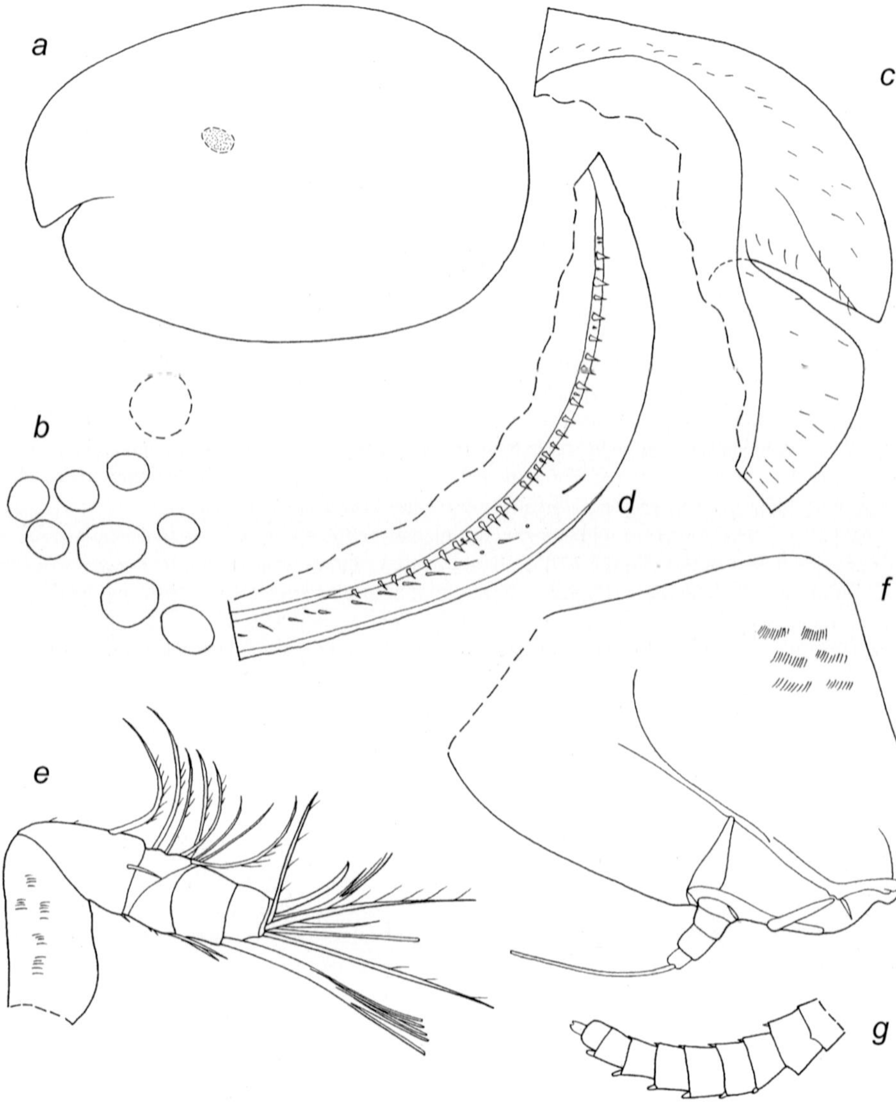

Figure 37. *Parasterope sigma* new species, holotype, BPBM-S 12840, adult female: *a,* complete specimen from left side, length 1.23 mm; *b,* central adductor muscles attachments left valve, iv; *c*, anterior left valve, iv; *d*, posterior right valve, iv; *e*, right 1st antenna, lv; *f*, protopod and endopod left 2nd antenna, mv; *g*, exopod right 2nd antenna, lv (nabs).

Seventh Limb (Figure 39*b*): Proximal group with 6 bristles, 3 on each side, each with 3 or 4 bells. Distal group with 6 bristles, 3 on each side, each with 2 to 4 bells. Terminus with opposing combs each with about 9 spinous teeth.

Furca (Figure 39*c,d*): Each lamella with 7 spinous claws followed by 2 backward pointing bristlelike claws. Total claws and bristles 9.

Bellonci Organ (Figure 38*f,g*): Elongate with rounded tip.

Eyes: Lateral eye with many ommatidia obscured by black pigment; 12 ommatidia visible

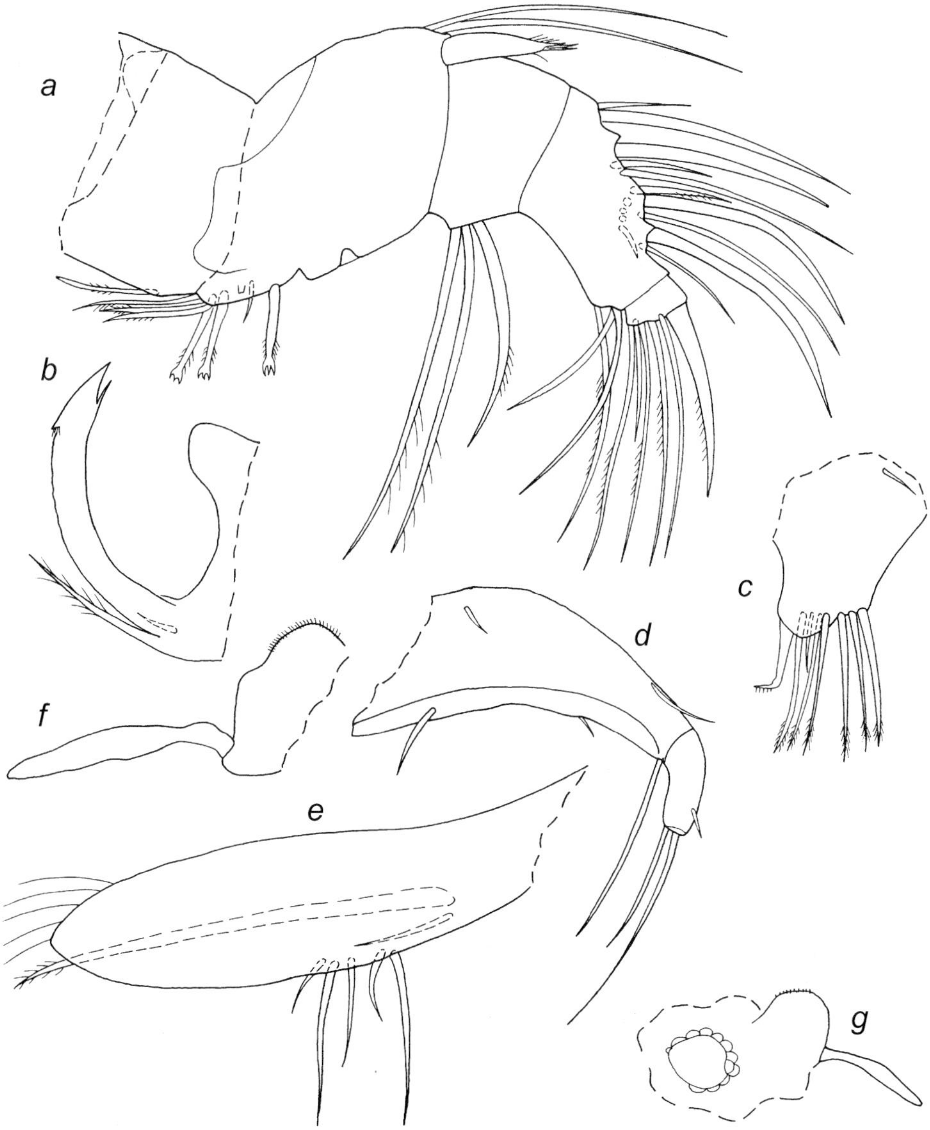

Figure 38. *Parasterope sigma* new species, holotype, BPBM-S 12840, adult female: *a,* right mandible, lv; *b,* coxa endite right mandible as seen through appendage (obscured), lv; *c*, proximal part right maxilla, lv; *d*, right maxilla, lv; *e*, comb right 5th limb, mv; *f*, medial eye and Bellonci Organ from left side; *g*, right lateral eye, medial eye, and Bellonci Organ.

around edge (Figures 37*a,* 38*g*). Medial eye about same size as lateral eye, unpigmented, with minute dorsal spines (Figure 38*f,g*).

Gills: Well developed with smooth distal edge.

Lips: Hirsute lobes (Figure 39*f*).

Anterior of Body: Without structures between 1st antenna and mandible.

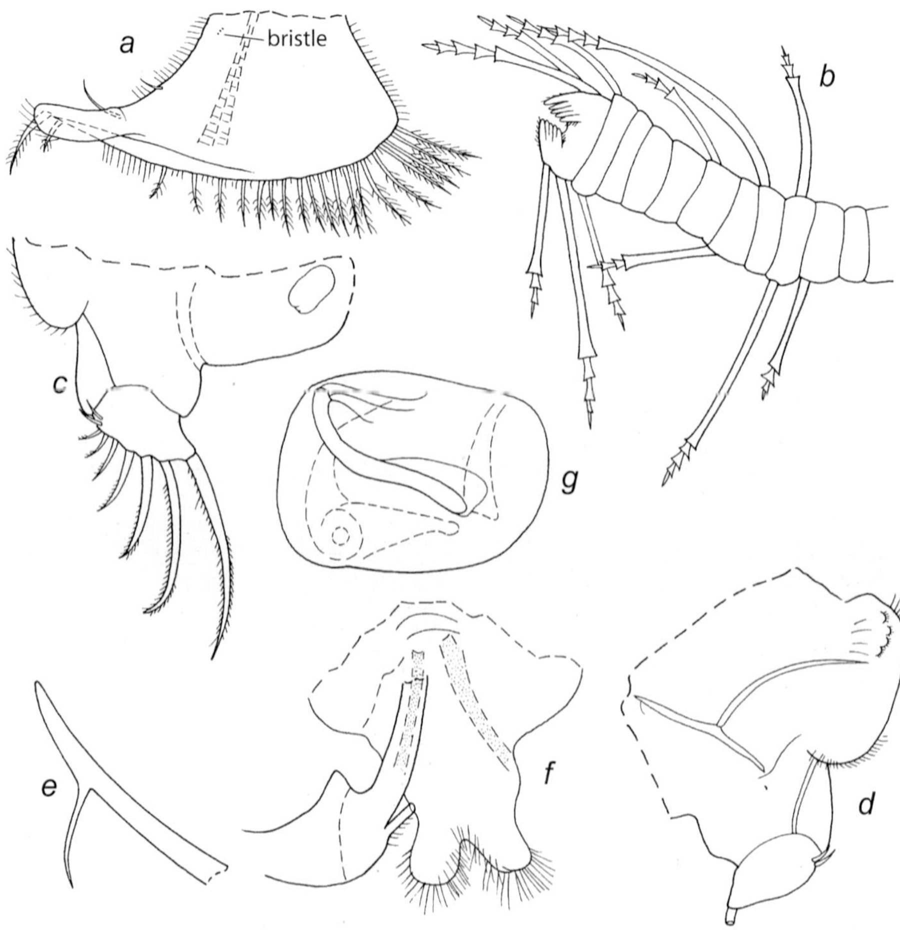

Figure 39. *Parasterope sigma* new species, holotype, BPBM-S 12840, adult female: *a,* left 6th limb, lv; *b,* 7th limb; *c*, posterior of body from right side showing right furcal lamella and genital organ (oval); *d*, posterior of body from left side (not all claws shown on furca); *e*, detail of anterior end of Y-Sclerite shown in "*d*"; *f*, posterior view of upper lip showing coxa endite of left mandible projecting into esophagus; *g*, genitalia on left side.

Posterior of Body (Figure 39*c,d*): Spinous with 4 spinous crescents on each side of posterodorsal corner.

Y-Sclerite (Figure 39*d,e*): With ventral projection near anterior end.

Genitalia (Figure 39*c, g*): Oval on each side of body with elongate and coiled internal structures.

Eggs: Holotype with small unextruded eggs.

Gut Content: Minute light-amber colored particulate matter.

Comparisons. The protopod of the 2nd antenna of the new species *P. sigma* differs from that of *P. pacifica* and *P. omega* in not having long spines on the distal dorsal corner and in having rows of minute medial spines near posterodorsal corner. The basal spines on the exopod of the 2nd antenna of *P. sigma* are smaller than those of *S. pacifica.* The dorsal margin of the basis of the mandibles of

P. sigma and *P. zamboangae* are without the midbristle present on *P. omega*. The posterodorsal corner of the body of *P. sigma* bears 4 spinous crescents compared to only 1 on *P. gamma*. The protopod of the 2nd antenna of the holotype of *P. zamboangae* was reexamined and it does not have the medial spines present on *P. sigma*. The comb of the 5th limb of *P. sigma* differs from that of *P. omega* in having 1 rather than 2 slender bristles just ventral to the base of the stout exopod bristle.

Parasterope zeta, new species

Figs. 40–44

Holotype. BPBM-S 12862, adult female on slides and in alcohol.

Type Locality. Sta 11, 22 Jan 2001, Kaiser's Channel, Waikīkī.

Paratypes. Waikīkī: Sta 8: USNM 1085960, partly dissected adult female on slide and in alcohol; Sta 10, BPBM-S 12863, 1 adult female, appendages and carapace in alcohol; Sta 11, BPBM-S 12864, 1 ovigerous female, appendages and valves in alcohol.

Non-types. Waikīkī: Sta 8, USNM 1085945, 1 ovigerous female and 1 juvenile (length 0.76 mm, height 0.60 mm) in alcohol, both unopened; Sta 11, USNM 1085961, 4 unopened adult females (2 ovigerous) in alcohol. French Frigate Shoals: Sta TC-13, USNM 1085962, ovigerous female on slide and in alcohol.

Distribution. Waikīkī: Sta 8, 10, 11. French Frigate Shoals: Sta TC-13.

Description of adult female (Figures 40–44). Carapace oval in lateral view (Figures 40*a*, 42*a*, 44*a*).

Infold: Rostral infold with total of about 30 short and long bristles. Anteroventral infold with 23 bristles and small bristle at inner end of incisure. Ventral infold with 11 bristles forming row. Infold of caudal process similar on both valves, with narrow ridge bearing about 27 flaplike bristles with 1 or 2 minute bristles between some pairs of flaplike bristle (Figure 40*b*); infold between ventral part of ridge and posteroventral corner of valve with 6 bristles forming row; small bristle present between dorsal end of ridge and valve edge.

Selvage: None observed in vicinity of incisure or along free margins.

Central Adductor Muscle Attachments (Figures 43*f*): 12 or 13 ovoid attachments.

Carapace Size (length, height in mm): BPBM-S 12862, 1.09, 0.69. BPBM-S 12864, 1.12, 0.71. USNM 1085961 (4 specimens): 1.12, 0.71; 1.10, 0.72; 1.16, 0.73; 1.12, 0.72. BPBM-S 12863, 1.10, 0.69. USNM 1085960, 1.16, 0.71. BPBM-S 12856, 1.10, 0.69. USNM 1085962, 1.10, 0.73.

First Antenna (Figures 40*c,d*, 42*b*): Article 1 bare. Article 2 with spinous dorsal bristle, distal lateral bristle with few minute marginal spines, distal spines forming lateral row in distal dorsal corner and in terminal ventral corner; few spines along ventral and dorsal margins. Article 3 with short ventral margin with small bristle and long dorsal margin with 6 bristles: 4 bristles with bases on edge of margin (3 proximal with long spines, 1 distal with short spines); 1 bristle lateral to base of 3rd edge bristle, and 1 lateral bristle with base lateral to 4th edge bristle (both lateral bristles with few long spines). Distinct diagonal suture separating 3rd and 4th articles and distinct concave suture separating 4th and 5th segments. 4th article with long terminal dorsal bristle with short spines, 2 short terminal bare ventral bristles (bristles shorter than length of 5th article), and few spines on ventral margin. Sensory bristle of 5th article long with 6 terminal filaments (5 ventral filaments longer than shown in Figure 40*c*). 6th article long with long terminal medial bristle with short spines. Article 7: a-bristle clawlike; b-bristle with 3 long filaments; c-bristle long with 4 short marginal filaments. Article 8: d-bristle absent; e-bristle well developed, bare with blunt tip; f-bristle bent dorsally, with 4 short filaments; g-bristle fairly long, with about 5 marginal filaments (4th filament much longer than others).

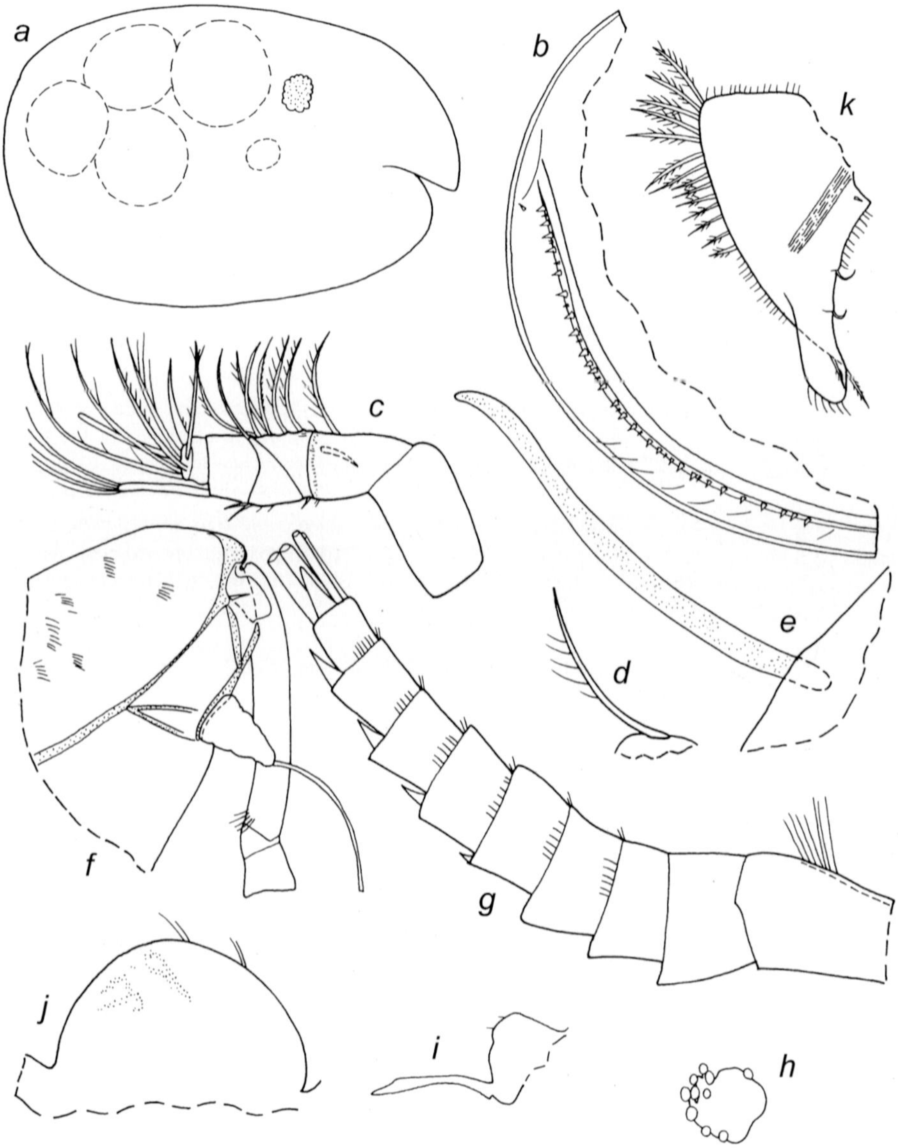

Figure 40. *Parasterope zeta,* new species, holotype, BPBM-S 12862, adult female: *a,* complete specimen from right side, length 1.09 mm; *b,* posterior left valve, iv; *c,* right 1st antenna, mv; *d,* foreign organism attached to dorsal bristle of 2nd article of left 1st antenna; *e,* pivot sclerite on posterodorsal edge of protopod of right 2nd antenna, lv; *f,* protopod and endopod left 2nd antenna, mv; *g,* exopod right 2nd antenna, lv (nabs); *h,* left lateral eye; *i,* medial eye and Bellonci Organ; *j,* medial eye; *k,* right 6th limb, lv.

Second Antenna: Protopod with small distal medial bristle, few distal short dorsal spines, and indistinct medial spines forming rows (Figures 40*f,* 42*c,* 44*b,d*). Endopod with 3 articles (fused or with sutures); terminal article with long filament (Figures 40*f,* 42*c,* 44*b,d*). Exopod (Figures 40*f,g,* 42*c,d*): article 1 with distal dorsal long spines. Bristle of article 2 reaching past 9th article; ventral margin with ventral spines and distal hairs. Bristle of article 3 long; ventral margin with spines and distal hairs; dorsal margin with long hairs. Bristle of articles 4 to 8 long and with natatory hairs. 9th

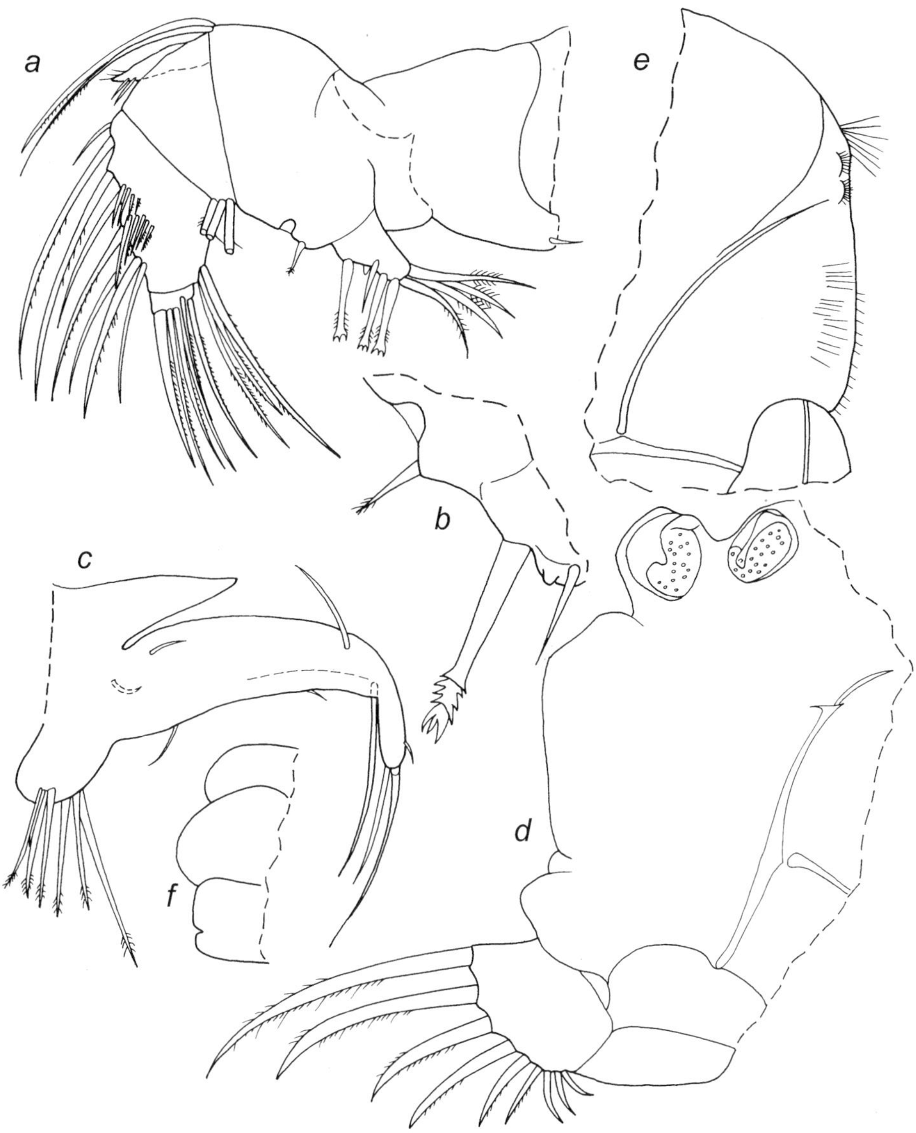

Figure 41. *Parasterope zeta,* new species, holotype, BPBM-S 12862, adult female: *a,* right mandible, mv; *b,* part basis and basis endite, detail from "*a*"; *c,* left maxilla, mv (nabs); *d,* posterior of body from left side showing left furcal lamella and genitalia; *e,* posterodorsal part of body from left side; *f,* posterior ends of ventral 3 gills on right side of body.

article with 4 bristles: 2 dorsal bristles short, bare or with short hairs; 2 ventral bristles long with natatory hairs, some also with few ventral spines. Articles 4 or 5 to 8 with basal spines longer on distal articles; spine of 8th article one-half to three-fourths length of 9th article. Article 9 with lateral spine three-fourths to full length of 9th article. Articles 2 or 4 to 8 with distal spines forming rows. Pivot sclerite of protopod slightly S-shaped (Figures 40*e,* 42*c*).

Mandible: (Figures 41*a,b,* 42*e,f,* 44*c*): Coxa endite broken off both limbs of BPBM-S 12862; small medial bristle at base of ventral branch (Figure 41*a*). Basis endite with 3 triaenid bristles, 1 fairly long dwarf bristle adjacent to small node, and 4 end bristles. Basis: ventral margin with short bristle either bare (Figure 42*f*) or with few minute distal spines (Figure 41*b*) proximal to U-shaped process; dorsal margin with 2 long terminal dorsal bristles with short spines. Exopod about same length as dorsal margin of article 1 of endopod, hirsute distally and with 2 small subterminal bristles. Endopod: article 1 with 3 ventral bristles (2 long with long spines, 1 shorter with short spines). Article 2: ventral margin with 3 dorsal terminal bristles with short spines; dorsal margin with stout a-, b-, c-, and d-bristles with short marginal spines (base of c-bristle similar in width to bases of b- and d-bristles) and 1 short bristle proximal to a-bristle; lateral side with a long bristle between b- and c-bristles and between c- and d-bristles; medial side with 2 to 4 cleaning bristles forming oblique row between b- and c-bristles, 5 or 6 cleaning bristles forming oblique row between c- and d-bristles, and fairly long bristle adjacent to base of d-bristle. Article 3 with stout dorsal clawlike bristle with small ventral teeth, 3 long terminal bristles with short marginal spines, 1 lateral bristle with short marginal spines just ventral to base of dorsal clawlike bristle, and 1 short medial bristle near ventral margin.

Maxilla (Figures 41*c*, 42*g*, 43*a*): Epipod triangular with few indistinct distal hairs. Endite I with 3 long bristles and 1 short bristle; endite II with 3 long bristles. Basis: ventral margin with short proximal backward oriented bristle, short distal bristle (not observed on USNM 1085962), and long terminal bristle; lateral side with short proximal bristle; medial side with short proximal bristle and long distal bristle, both near dorsal margin. Endopod: article 1 with small alpha-bristle and long beta-bristle; article 2 with long terminal bristle reaching past beta-bristle.

Fifth Limb (Figure 43*b*): Comb with stout spinous exopod bristle; 1 or 2 slender bristles with bases just ventral to base of exopod bristle. Several bristles closer to ventral margin.

Sixth Limb (Figures 40*k*, 43*c*): Small medial bristle at anterior proximal corner. Endites I and II each with single bristle. Anterior corner of ventral margin with 2 spinous bristles; ventral margin with 14 spinous bristles (bristles along posteroventral corner longer than others). Lateral flap with long hairs but no bristles. Anterior, ventral, and posterior margins and medial surface hirsute.

Seventh Limb: Proximal group with 6 bristles, 3 on each side, each with 3 or 4 bells. Distal group with 6 bristles, 3 on each side, each with 3 to 4 bells. Terminus with opposing combs, each with 7 spinous teeth.

Furca (Figures 41*d*, 43*d,e*): Each lamella with 6 claws and 3 bristles; claws with spines along posterior concave margins; claws 1 to 3 or 4 with distal hairs along anterior convex margins. Right lamella anterior to left by width of base of claw 1.

Bellonci Organ (Figures 40*i,j*, 44*d*): Elongate with rounded tip.

Eyes: Lateral eye with many ommatidia (at least 15, most obscured by black pigment) (Figures 40*a,h*, 42*a*, 44*a,d*). Medial eye about same size as lateral eye, with slight amount of brown pigment and few dorsal hairs visible at high magnification (Figures. 40*i,j*, 44*d*).

Gills (Figure 41*f*): Well developed with evenly rounded tips.

Lips: Partly obscured, may have 1 or 2 anterior spines.

Posterior of Body (Figures 41*e*, 43*e*): Posterodorsal corner evenly rounded, with long spines, and with 2 spinous crescent like processes on each side.

Y-Sclerite (Figures 41*d*, 43*e*): With pointed ventral process at anterior end.

Genitalia (Figures 41*d*, 43*c*): Oval on each side with internal structures.

Eggs in Marsupia (Figures 40*a*, 42*a*): BPBM-S 12862–4; BPBM-S 12864–6; USNM 1085961 (2 specimens) – 2, 9. USNM 1085960–9. BPBM-S 12856–6. USNM 1085962–5.

Gut Content: Amber colored unidentified particulate matter.

Foreign organisms: BPBM-S 12862 with few segmented unbranched filaments attached to endopod of left 2nd antenna, left 1st antenna (Figure 40*d*), and both mandibles. BPBM-S 12864 with similar filament on endopod of right 2nd antenna and several filaments on exopod of left 2nd antenna. Posterior edge of carapace of BPBM-S 12864 with several stemmed cups.

Comparisons. The mandible of *P. zeta* differs from that of *P. omega* in not having a dorsal midbristle on the basis. The mandible of *P. zeta* differs from those of *P. pacifica* and *P. zamboangae* in having a non-triaenid bristle with or without few weakly developed minute spines rather than a well

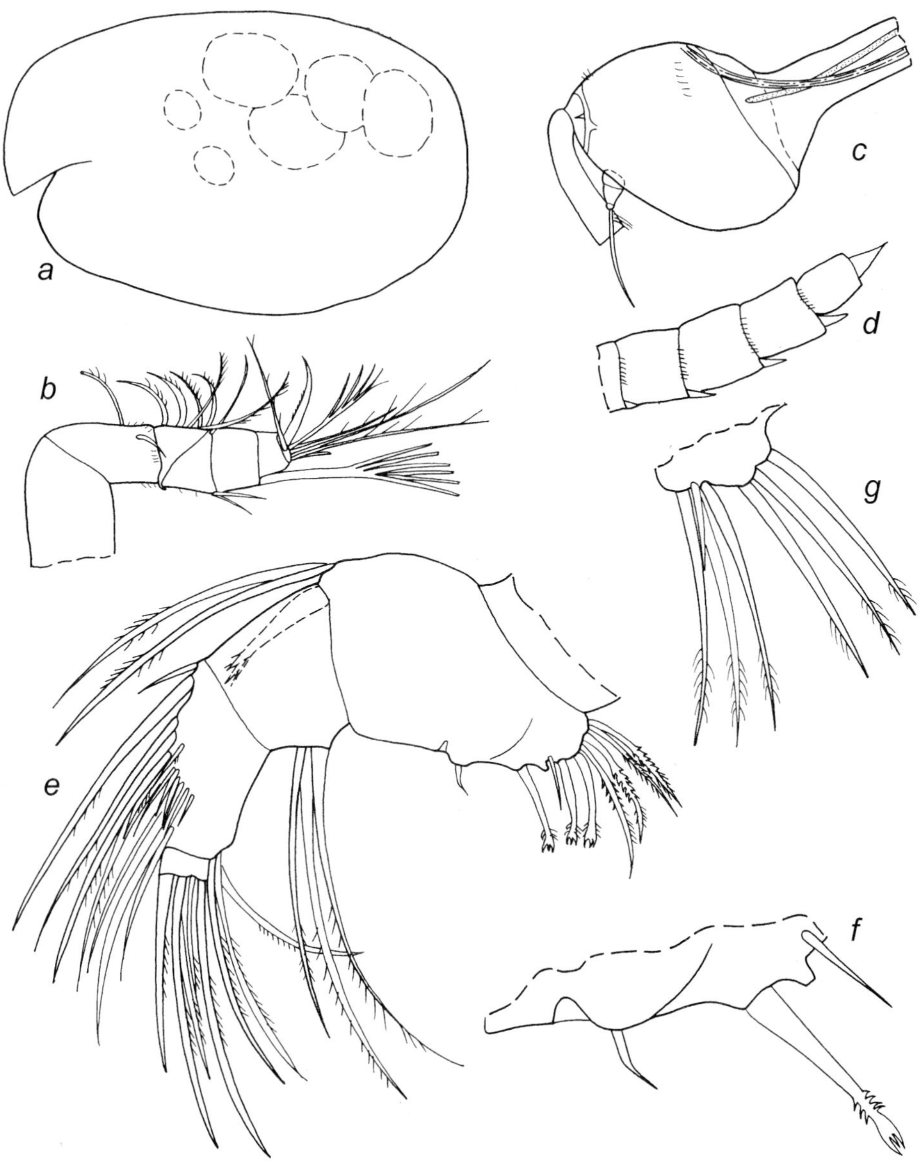

Figure 42. *Parasterope zeta,* new species, non-type, USNM 1085962, adult female: *a,* complete specimen from left side, length 1.10 mm; *b,* right 1st antenna, lv; *c*, protopod, endopod, 1st exopod article right 2nd antenna, mv; *d*, distal articles of exopod left 2nd antenna, lv (nabs); *e*, right mandible, mv; *f*, part basis and basis endite, detail from "*e*"; *g*, endites I (left) and II left maxilla, mv.

developed triaenid bristle on the ventral margin of the basis. A bristle is not present on the ventral margin of the basis of *P. sigma*. A bristle is also absent from the ventral margin of the basis of *P. gamma*. The latter species also differs from *P. zeta* in not having basal spines on the exopod of the 2nd antenna.

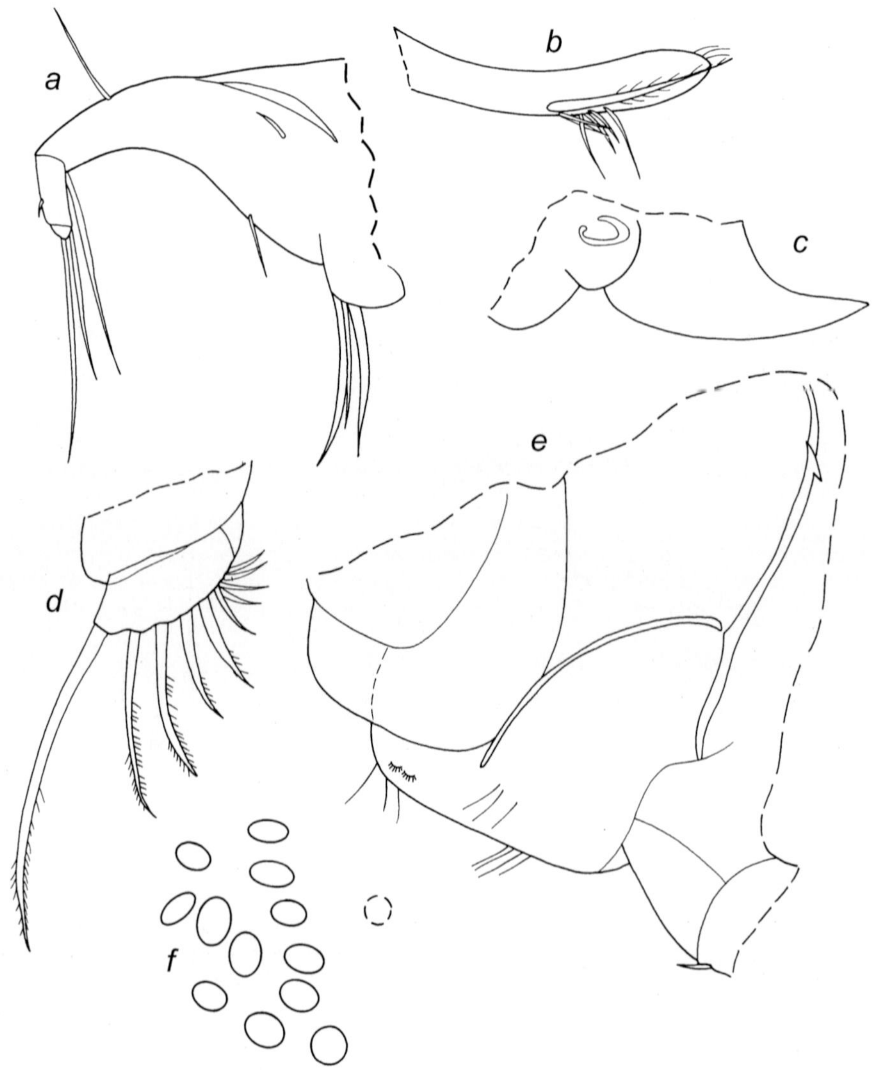

Figure 43. *Parasterope zeta,* new species, non-type, USNM 1085962, adult female: *a,* right maxilla, mv (nabs); *b,* comb right 5th limb, lv (nabs); *c*, genitalia and outline of right 6th limb (nabs); *d*, left furcal lamella, mv; *e*, posterior of body from right side, not all claws of furca shown. Paratype, BPBM-S 12864, adult female: *f*, central adductor muscle attachments of right valve, ov.

Parasterope theta, new species

Figs. 45–47

Holotype. BPBM-S 12865, adult male on 2 slides and in alcohol.

Type Locality. Sta 11, 22 Jan 2001, Kaiser's Channel, Waikīkī.

Paratypes. None.

Distribution. Waikīkī, Sta 11.

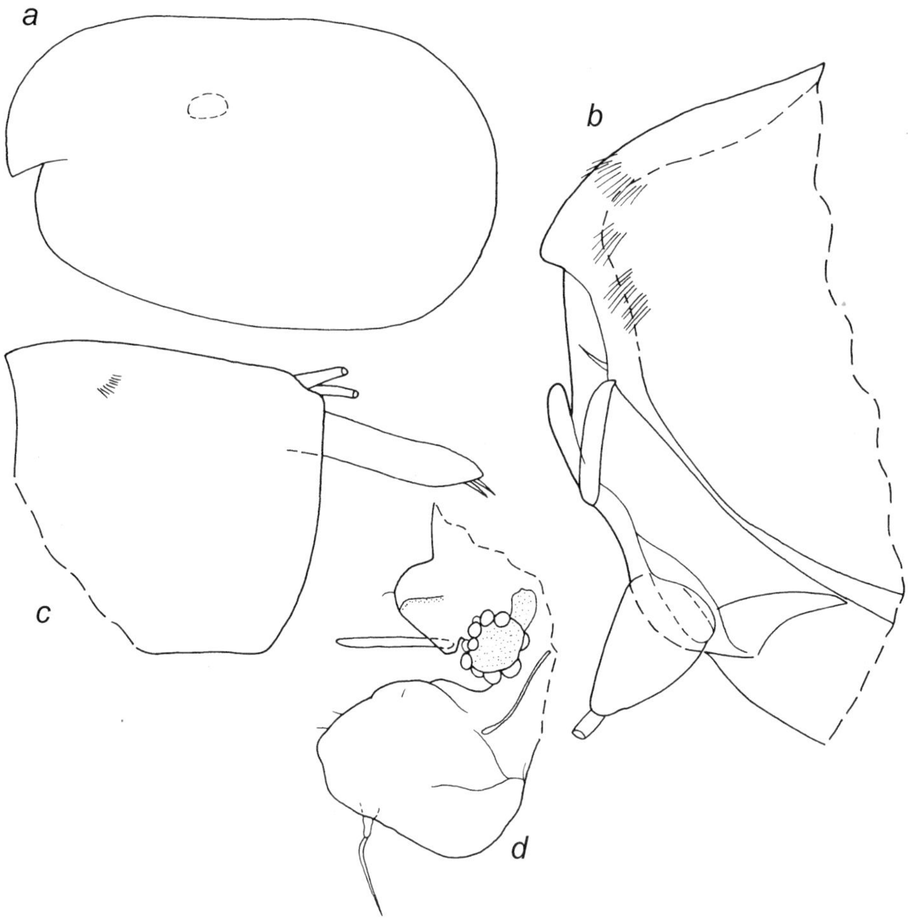

Figure 44. *Parasterope zeta,* new species, paratype, USNM 1085960, adult female: *a,* complete specimen from left side, length 1.16 mm; *b,* distal protopod and endopod right 2nd antenna, mv; *c*, basis and exopod right mandible, lv. Paratype, BPBM-S 12863, adult female: *d*, anterior of body from left side showing medial eye and Bellonci Organ, left lateral eye, protopod and endopod left 2nd antenna.

Description of adult male (Figures 45–47). Carapace elliptical in lateral view and with oblique dorsal margin near posterior end (Figure 45*a*). Few outer bristles present along posterior margin (Figure 45*a,d*).

Infold: Infold of rostrum with about 20 bristles (Figure 45*c*). Small bristle at inner end of incisure. Anteroventral infold with about 20 bristles (Figure 45*b*). Ventral infold with about 9 bristles. Posteroventral infold with broad band at inner edge of infold with about 20 flaplike bristles with 1 or 2 small bristles between some pairs of flaplike bristles; 4 long bristles between ventral end of broad band and posteroventral corner of valve (Figure 45*d*).

Selvage: Narrow selvage with long hairs along ventral edge of incisure (Figure 45*c*). None observed along free margins.

Central Adductor Muscle Attachments: Consisting of about 11 ovoid attachments.

Carapace Size (length, height in mm): BPBM-S 12865, 0.92, 0.54.

First Antenna (Figure 45*e*): Article 1 bare. Article 2 with distal dorsal bristle with long spines and shorter distal lateral bristle. Article 3 with short ventral margin with small bristle, and long dorsal margin with 6 bristles (4 edge bristles (3 proximal with long spines, 1 distal), 1 slightly lateral bristle (with few long spines) between 2nd and 3rd edge bristles, and 1 slightly lateral bristle adjacent to 4th edge bristle). Oblique suture between 3rd and 4th article; article 4 with 1 long dorsal bristle and 2 short ventral bristles. Article 5 short with stout ventral sensory bristle with abundant long filaments. Article 6 short, with long medial bristle near dorsal margin, and small dorsal indentation about midlength. 7th article: a-bristle clawlike with minute spines along dorsal edge; b-bristle longer than a-bristle, with 4 marginal filaments; c-bristle long with 10 short filaments. 8th article: d-bristle absent; e-bristle longer than a-bristle, bare with blunt tip; f-bristle longer than b-bristle, with about 5 short marginal filaments; g-bristle broken on both limbs of specimen, remaining part longer than f-bristle but shorter than g-bristle, with 6 marginal filaments.

Second Antenna: Protopod with small distal medial bristle, otherwise bare (Figure 46*a, c*). Endopod with 3 articles (Figure 46*a–c*): article 1 bare; article 2 with 3 dorsal bristles near midlength; article 3 reflexed on article 2, with proximal filament and tip with 4 rounded teeth (Figure 46*b*). Exopod with 9 articles: article 1 with small triangular lateral terminal process; basal spines absent, but articles 2 to 8 with lateral partly internal sclerites (Figure 46*d*) that superficially resemble lateral spines. Distal articles with spines along parts of terminal edge; Articles 2 to 8 with long dorsal spines (Figure 46*d*). Article 9 with pointed lateral spine (Figure 46*d*). Article 2 with fairly long bristle with spines or natatory hairs; articles 3 to 8 with long bristles with natatory hairs, some with proximal ventral spines; 9th article with 4 bristles: dorsal bristle short with short proximal spines and long natatory hairs near tip; bristle adjacent to dorsal bristle medium in length and with natatory hairs; remaining 2 bristles long and with natatory hairs.

Mandible: Coxa endite: ventral branch with 4 rows of spines and 3 small terminal spines (Figure 46*e*); dorsal branch with ventral nodes, spine at tip, and subterminal dorsal bristle; slender medial bristle near base of ventral branch. Basis endite with 3 spinous end bristles, 2 triaenid bristles with 3 pairs of spines proximal to terminal pair, and 1 fairly long dwarf bristle adjacent to peg (Figures 45*f*, 46*f*). Basis: ventral margin with small plain bristle proximal to U-shaped process; dorsal margin with 2 long subterminal bristles. Exopod almost reaching distal margin of 1st endopod article, with 2 small subterminal bristles and spinous tip. Endopod: article 1 with 3 ventral bristles (2 with long spines, shorter bristle with small spines). Article 2: ventral margin with 3 terminal ventral bristles with short marginal spines; dorsal margin with stout a-, b-, c-, and d-bristles, and 2 short slender proximal bristles; lateral side of dorsal margin with long slender bristle between b- and c-bristles and c- and d-bristles; medial side near dorsal margin with 2 spinous cleaning bristles between b- and c-bristles, oblique row of 6 cleaning bristles adjacent to c-bristle, and 1 long slender bristle adjacent to d-bristle. Article 3 with stout dorsal clawlike bristle with ventral spines, 3 long end bristles, 1 short medial bristle near ventral margin, and 1 long lateral bristle just ventral to base of dorsal clawlike bristle.

Maxilla (Figure 47*a*): Epipod triangular with pointed tip (not shown). Endite I with 4 bristles (3 long and 1 short); endite II with 3 long bristles. Basis: dorsal margin with indistinct proximal lateral bristle and fairly long distal dorsal bristle; ventral margin with proximal backward-pointing bristle and long spinous terminal bristle. Endopod: article 1 with small alpha-bristle and long beta-bristle; article 2 with terminal bristle reaching past beta-bristle.

Fifth Limb (Figure 45*g*): Comb with stout spinous exopod bristle reaching past tip of comb; 1 slender bristle present just ventral to base of exopod bristle; 5 bristles forming row distal to base of exopod bristle and close to ventral margin.

Sixth Limb (Figure 47*b*): Anterior margin with upper and lower endite bristles. Anterior corner of ventral margin with 2 bristles (posterior bristle about one-half length of anterior bristle). Posterior one-half of ventral margin with 12 spinous bristles (bristles longer at posteroventral corner); anterior, ventral, and posterior edges of limb hirsute. Narrow lateral flap hirsute but without bristles.

Seventh Limb: Proximal group with 6 bristles, 3 on each side, each with 3 or 4 bells. Distal group with 6 bristles, 3 on each side, each with 2 to 4 bells. Terminus with opposing combs, each with about 6 teeth.

Furca (Figure 47*c*): Each lamella with 6 claws and 2 posterior bristlelike claws; Claw 1 with-

Figure 45. *Parasterope theta,* new species, holotype, BPBM-S 12865, adult male: *a,* complete specimen with projecting copulatory organ, length 0.92 mm; *b, c,* anterior left valve, iv; *d,* posterior left valve, iv; *e,* left 1st antenna, lv; *f,* left mandible, lv; *g,* part comb right 5th limb, lv (nabs); *h,* medial eye and Bellonci Organ; *i,* right lateral eye.

out spines along posterior concave edge and slender distal spines along anterior edge. Claws 2 to 5 with spines along posterior concave edge; claws 2 and 3 with slender distal spines along anterior convex edge. Right lamella anterior to left by width of base of claw 1.

Bellonci Organ (Figure 45*h*): narrow proximally, broad at midlength, then tapers to broadly rounded tip.

Eyes: Lateral eye with 19 ommatidia, many of these obscured by black pigment; about 9 visible around edge of eye (Figure 45*a,i*). Medial eye unpigmented, about same size as lateral eye, bare (Figure 45*h*).

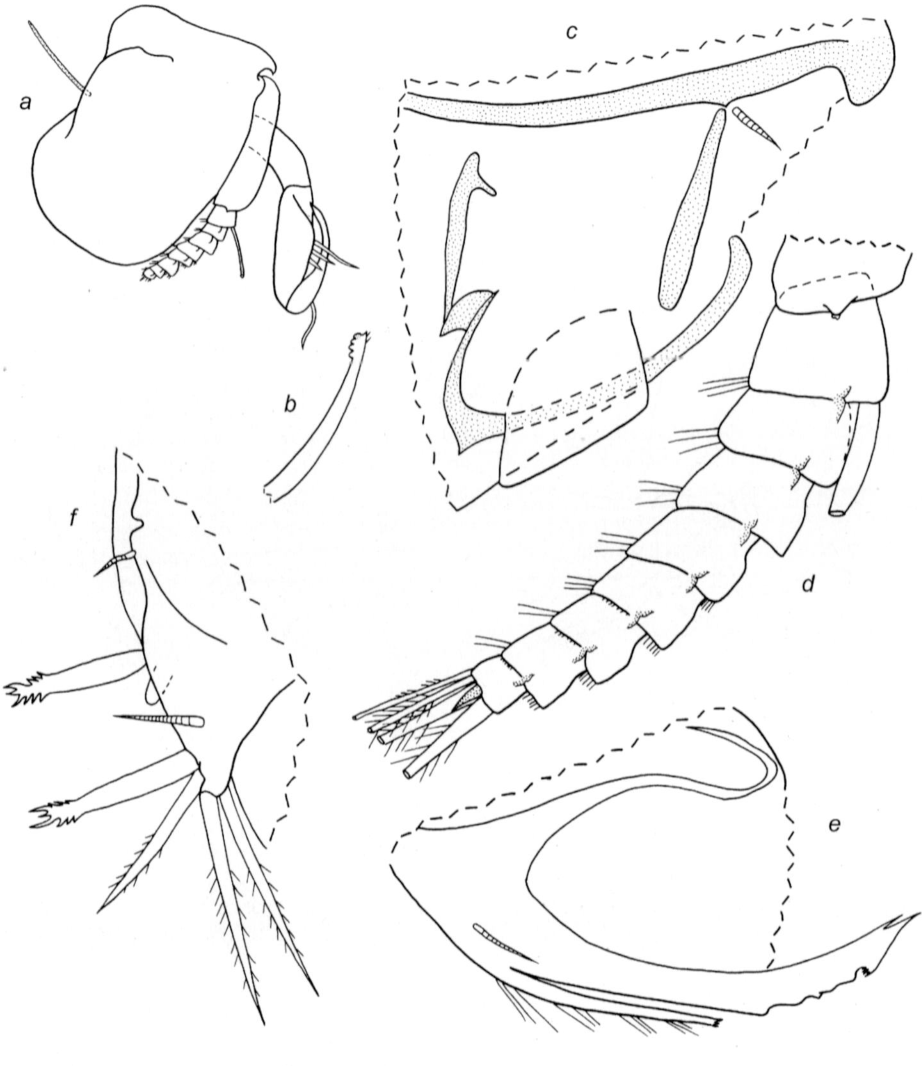

Figure 46. *Parasterope theta,* new species, holotype, BPBM-S 12865, adult male: *a,* right 2nd antenna, lv (nabs); *b,* 3rd endopod article, left 2nd antenna, mv; *c*, distal protopod left 2nd antenna, mv; *d*, exopod right 2nd antenna, lv (nabs); *e*, coxa endite right mandible, mv; *f*, basis right mandible, mv.

Gills (Figure 47*c*): Well developed with smooth rounded tips.

Lips: Not observed.

Anterior of Body: Small triangular protuberance between 1st antenna and mandible.

Posterior of Body: With smooth rounded posterodorsal corner. Without spines or lateral crescents.

Y-Sclerite (Figure 47*c,e*): With ventral pointed process near anterior tip.

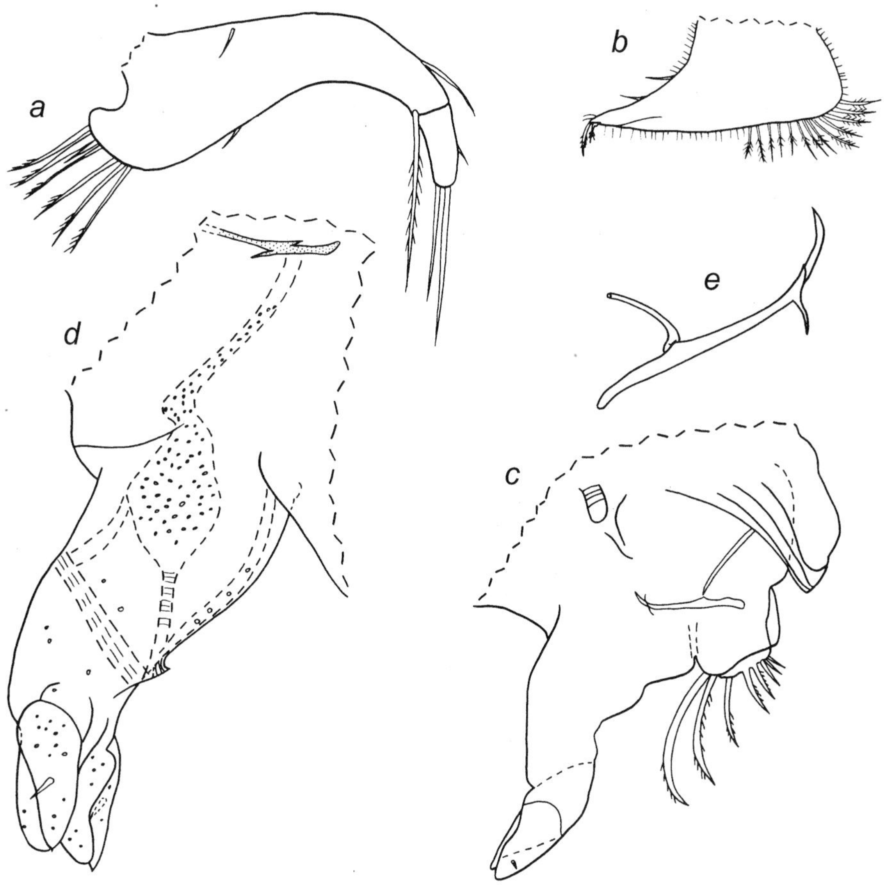

Figure 47. *Parasterope theta,* new species, holotype, BPBM-S 12865, adult male: *a,* right maxilla, lv (nabs); *b,* left 6th limb, lv; *c*, posterior of body from left side; *d*, copulatory organs from right side; *e*, right Y-Sclerite and part of right girdle.

Genitalia (Figure 47*c,d*): Copulatory organ with paired lobes on each side (4 lobes); outer lobe of each pair with small lateral bristle or spine and minute process at tip; small flared anterior process proximal to base of lobes may represent penis (Figure 47*d*).

Gut Content: Light amber particulate matter.

Remarks. The holotype of this species is the only adult male of *Parasterope* in the collections from Hawai'i. Unable to relate the male with certainty to any of the species of *Parasterope* described herein from Hawai'i, which are known only from females, *P. theta* is proposed as a new species.

Comparisons. The unique male of *P. theta* was present in the same sample at Sta 11 as *P. zeta* described above. They resemble each other in both having a plain bristle on the ventral margin of the basis of the mandible proximal to the U-shaped process. The exopods of the 2nd antennae differ in that articles 4 to 8 of *P. zeta* have basal spines, whereas those articles of *P. theta* have none. It is not unusual for male basal spines to be smaller than those of the female of the same species, but not to the extent of being present on the female and absent on the male. The length of the adult male carapace of *P. theta* is 0.92 mm compared to 1.09–1.06 mm for the female *P. zeta*, but the difference would not be too unusual if they are conspecific. Basal spines are absent on the exopod of the female 2nd antenna of *P. gamma*, but the mandible is without the plain ventral bristle on the mandible basis present on *P. theta*. The comb of the 5th limb of *P. theta* differs from that of *P. omega* in having 1 rather than 2 slender bristles just ventral to the base of the stout exopod bristle.

Parasterope iota, new species

Figs. 48–50

Holotype. BPBM-S 12866, adult female on 3 slides and in alcohol.

Type locality. Sta 10, 22 Jan 2001, Kapua Channel, Waikīkī.

Paratypes. Waikīkī: Sta 7, Canoes, USNM 1085963, partly dissected adult female in alcohol.

Non-type. Waikīkī: Sta 7, Canoes, BPBM-S 12857, undissected adult female in alcohol.

Distribution. Waikīkī: Kapua Channel, Canoes.

Description of adult female (Figures 48–50). Carapace elliptical in lateral view (Figures 48*a*, 50*c*).

Infold: Infold of rostrum (including row of 9 dorsal bristles in anterodorsal corner of valve) with about 31 bristles. Small bristle at inner end of incisure. Anteroventral infold with about 26 bristles. Ventral infold with about 13 bristles. Posteroventral infold with broad band at inner edge of infold with about 29 flaplike bristles with 1 or 2 small bristles between some pairs of flaplike bristles; 8 or 9 long bristles between ventral end of broad band and posteroventral corner of valve. Posterior infold of left and right valves similar.

Selvage: None observed in vicinity of incisure or along free margins.

Central Adductor Muscle Attachments (Figure 48*b*): Consisting of about 11 ovoid attachments.

Carapace Size (length, height in mm): BPBM-S 12866, 1.18, 0.83. USNM 1085963, 1.20, 0.79. BPBM-S 12857, 1.24, 0.83.

First Antenna (Figures 48*c*, 49*a*): Article 1 with long lateral and medial distal spines in anterodorsal corner (Figure 49*a*). Article 2 with distal dorsal bristle with long spines, shorter distal lateral bristle, 2 rows of distal ventral spines, few proximal dorsal spines, and row of lateral spines along terminal edge near dorsal margin. Article 3 with short ventral margin with small bristle, and long dorsal margin with 6 (left limb) or 7 (right limb) bristles: 4 (5 on left limb) edge bristles with long spines (4th bristle with few spines), 1 slightly lateral bristle (with few long spines) between 2nd and 3rd edge bristles, and 1 slightly lateral bristle (with long spines) adjacent to 4th edge bristle. Oblique suture between 3rd and 4th articles; article 4 with 1 long dorsal bristle with short spines along posterior edge, 2 short ventral bristles, and few spines along ventral margin. Article 5 with long stout ventral sensory bristle with 6 long terminal filaments. Article 6 with long medial bristle (with short marginal spines) near ventral margin. 7th article: a-bristle clawlike, bare; b-bristle longer than a-bristle, with 1 short proximal and 3 long distal marginal filaments; c-bristle long with 5 short filaments. 8th article: d-bristle absent; e-bristle longer than a-bristle, bare with blunt tip; f-bristle bent dorsally, slightly longer than b-bristle, with 3 short marginal filaments; g-bristle almost as long as c-bristle, with 5 marginal filaments.

Second Antenna (Figure 49*b,c*): Protopod with small distal medial bristle, otherwise bare. Endopod with 3 articles: articles 1 and 2 bare; article 3 with long terminal filament. Exopod: bristle of article 1 reaching past article 9, with slender, ventral, short, hairlike spines; bristles of articles 3 to 8 with fairly stout proximal ventral spines and distal natatory hairs; 9th article with 4 bristles (2 short dorsal bristles (longer bristle with few hairs), 1 middle bristle longer and with few natatory hairs, and long ventral bristle with proximal ventral spines and distal natatory hairs). Article 1 with small triangular lateral terminal process at midwidth; articles 4 to 8 with basal spines with rounded

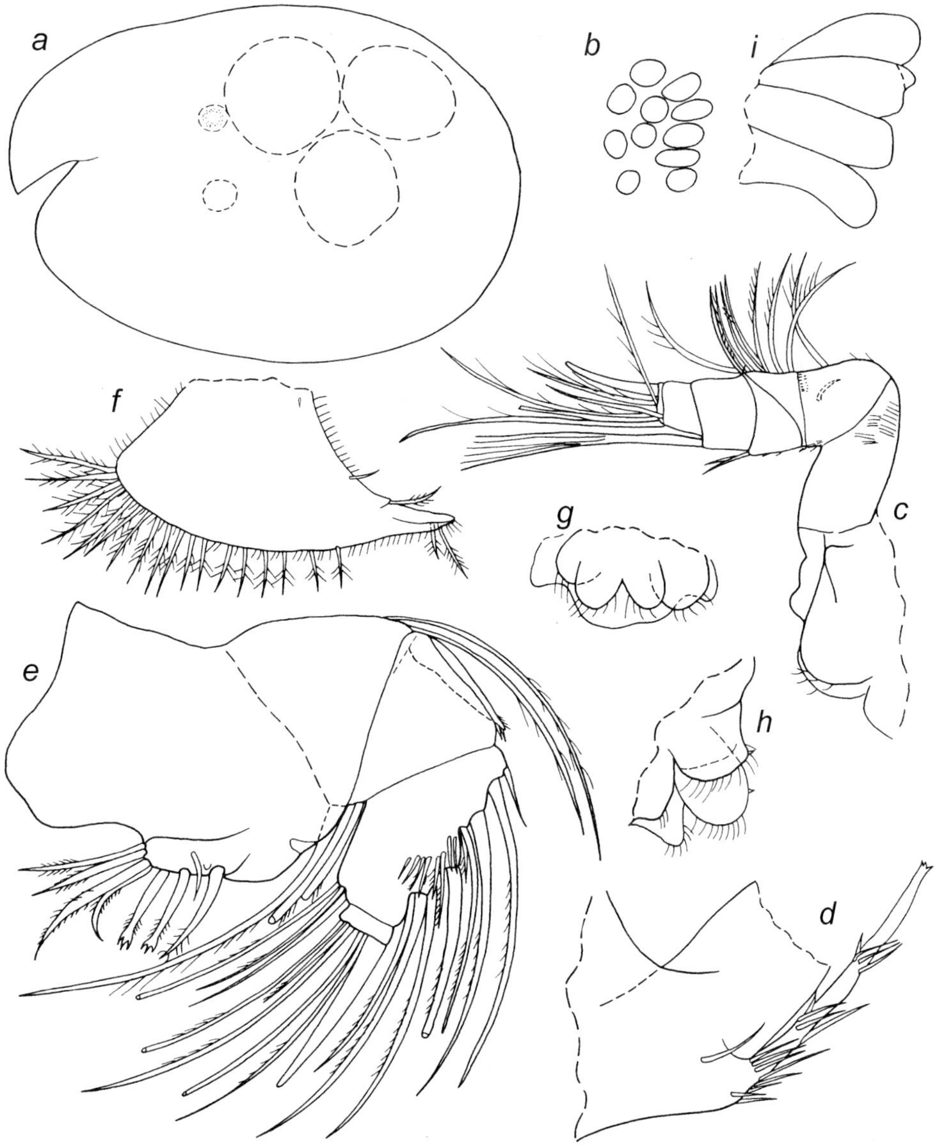

Figure 48. *Parasterope iota,* new species, holotype, BPBM-S 12866, adult female: *a,* complete specimen from left side, length 1.18 mm; *b,* central adductor muscle attachments left valve, ov; *c*, anterior of body from left side and right 1st antenna, mv; *d*, ventral branch coxa endite right mandible, mv; *e*, left mandible, mv; *f*, right 6th limb, lv; *g*, *h*, upper and lower lips from anterior (top) and right side; *i*, 4 gills from left side.

tips; spine of 8th article about one-half length of article 9. Article 9 with lateral spine (with rounded tip) about same length as article 9. Articles 4 to 8 with thin spines forming row along distal edges and stouter terminal spines on dorsal corner.

Mandible (Figure 48*d,e*): Coxa endite: ventral branch with 3 or 4 rows of spines and 3 terminal spines; dorsal branch broken off both limbs of holotype; small slender medial bristle near base of ventral branch (Figure 48*d*). Basis endite with 4 spinous end bristles, 3 triaenid bristles with 3 or 4 pairs of spines proximal to terminal pair, and 1 fairly long dwarf bristle adjacent to peg (Figure

48*e*). Basis: ventral margin with U-shaped process; dorsal margin with 2 long spinous subterminal bristles. Exopod reaching distal margin of 1st endopod article, with 2 small subterminal bristles and spinous tip. Endopod: article 1 with 3 ventral bristles (2 with long spines, shorter bristle with small spines). Article 2: ventral margin with 3 terminal ventral bristles with short marginal spines; dorsal margin with stout spinous a-, b-, c-, and d-bristles (base of c-bristle stouter than others), and 1 short slender proximal bristle; lateral side of dorsal margin with long slender spinous bristle between b- and c-bristles and c- and d-bristles; medial side near dorsal margin with 3 spinous cleaning bristles between b- and c-bristles, oblique row of 6 cleaning bristles adjacent to c-bristle, and 1 long spinous slender bristle adjacent to d-bristle. Article 3 with stout dorsal claw, 3 long spinous end bristles, 1 short spinous medial bristle near ventral margin, and 1 long spinous lateral bristle just ventral to base of dorsal claw. USNM 1085963 with 4 triaenid bristles on basis endite and endopodial article 2 with 1 cleaning bristle between b- and c-bristles, and 5 cleaning bristles between c- and d-bristles.

Maxilla (Figure 50*a*): Epipod triangular with pointed tip. Endite I with 3 long bristles and 1 short bristle; endite II with 3 long bristles. Basis: lateral side with proximal bristle near midheight; dorsal margin with proximal medial bristle and fairly long distal bristle; ventral margin with proximal backward-pointing bristle, small distal bristle and long spinous terminal bristle. Endopod: article 1 with small alpha-bristle and long beta-bristle; article 2 with terminal bristle about twice length of beta-bristle.

Fifth Limb (Figure 49*d*): Comb with stout spinous exopod bristle reaching past tip of comb; 1 slender bristle present just ventral to base of exopod bristle; 5 bristles just distal to base of exopod bristle and close to ventral margin; 1 proximal bristle near ventral margin.

Sixth Limb (Figure 48*f*): Anterior margin with upper and lower endite bristles (lower bristle fairly long and with marginal hairs). Anterior corner of ventral margin with 2 bristles (posterior bristle about one-half length of anterior bristle). Anterior corner separated by space from 18 spinous bristles (bristles longer at posteroventral corner). Anterior, ventral, and posterior edges of limb hirsute. Narrow lateral flap hirsute but without bristles. Medial surface spinous and with minute bristle in proximal anterior corner.

Seventh Limb: Proximal group with 6 bristles, 3 on each side, each with 3 or 4 bells. Distal group with 6 bristles, 3 on each side, each with 3 or 4 bells. Terminus with opposing combs, each with 9 or 10 spinous teeth.

Furca (Figure 50*b,d*): Each lamella with 6 claws and 3 posterior bristlelike claws. Claws 1 to 6 with spines along posterior concave edge; claws 1 to 4 with slender distal spines along anterior convex edge. Right lamella anterior to left by width of base of claw 1.

Bellonci Organ (Figure 49*e*): Elongate with broadly rounded tip.

Eyes: Lateral eye with many ommatidia, many of these obscured by black pigment (Figures 48*a*, 49*f*, 50c). Medial eye with small amount of light amber-colored pigment, about same size as lateral eye, with short dorsal spines (Figure 49*e*).

Gills (Figures 48*i*, 50*d*): Well developed with rounded tips.

Lips (Figure 48*g,h*): Paired anterior lobes of upper lip hirsute and with anterior spine; lower lip with hirsute paired lobes.

Anterior of Body (Figure 48*c*): Undulate.

Posterior of Body: With spines along margin proximal to posterior end of girdle (Figure 50*d*). Part of posterodorsal margin of holotype dorsal to posterior end of girdle of BPBM-S 12866 broken off during dissection and details obscured. USNM 1085963 with 4 spinous crescents dorsal to posterior end of girdle (Figure 50*d*).

Y-Sclerite (Figure 50*b*): With ventral pointed process near anterior tip.

Genitalia: Not examined.

Eggs in Marsupium (Figure 48*a*, 50*c*): BPBM-S 12866 with 9 eggs; length of 1 egg 0.30 mm (Figure 49*g*, left); illustrated egg with small foreign ovoid attached (Figure 49*g*, right). USNM 1085963 with 3 eggs. BPBM-S 12857 with 6 eggs.

Gut Content: Light amber particulate matter.

Foreign Organisms: One egg of BPBM-S 12866 with attached small unidentified ovoid (length

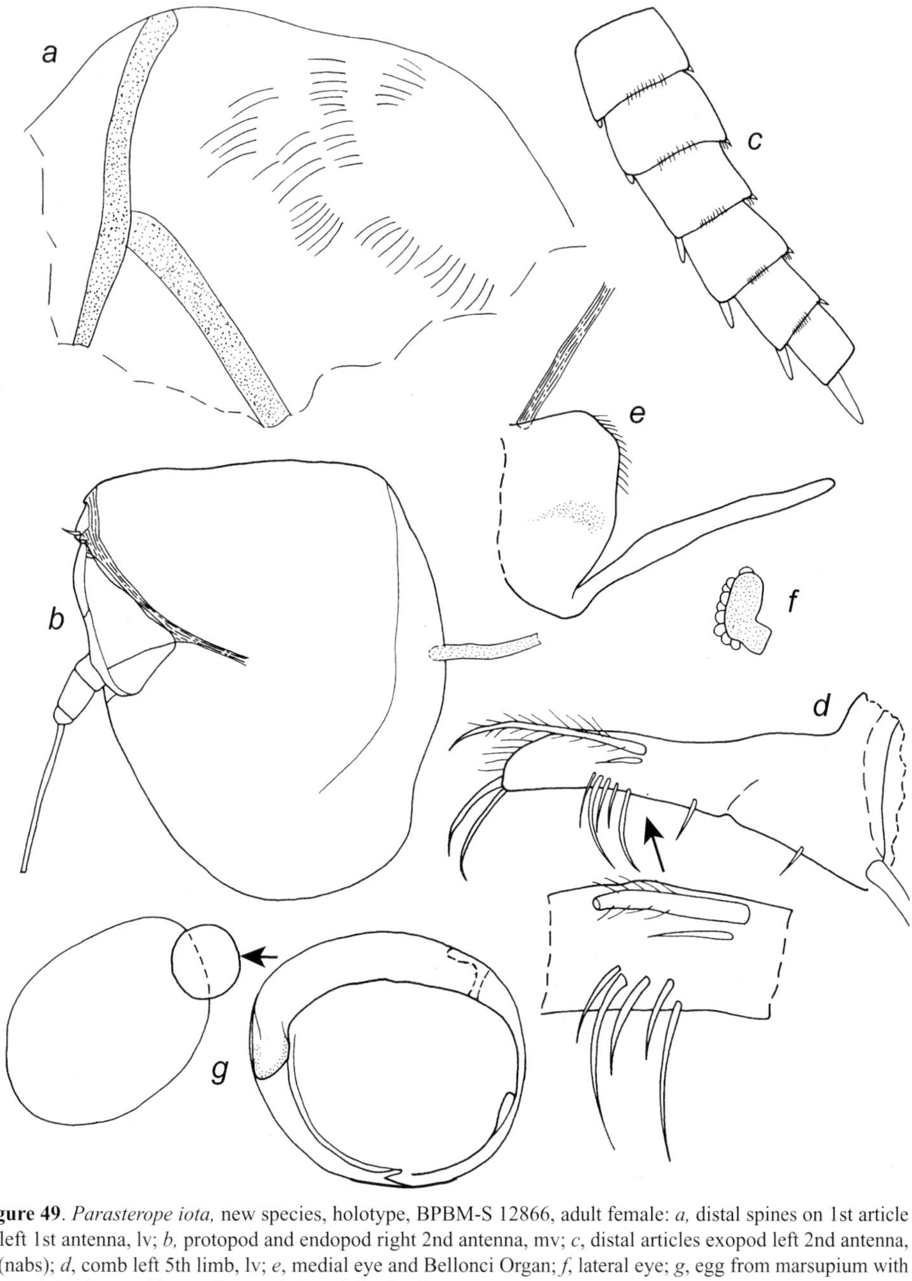

Figure 49. *Parasterope iota,* new species, holotype, BPBM-S 12866, adult female: *a,* distal spines on 1st article of left 1st antenna, lv; *b,* protopod and endopod right 2nd antenna, mv; *c,* distal articles exopod left 2nd antenna, lv (nabs); *d,* comb left 5th limb, lv; *e,* medial eye and Bellonci Organ; *f,* lateral eye; *g,* egg from marsupium with attached foreign ovoid organism on left, and detail of organism on right.

0.085 mm) with internal structures (Figure 49*g,* right). Exopod of right 2nd antenna of BPBM-S 12866 with segmented elongate filamentlike organism on bristle of 2nd article.

Comparisons: The new species *P. iota,* is close to *P. pacifica*. The protopod of the 2nd antenna of *P. pacifica* bears distal dorsal spines absent on *P. iota,*, and the exopod of the 2nd antenna of *P. paci-*

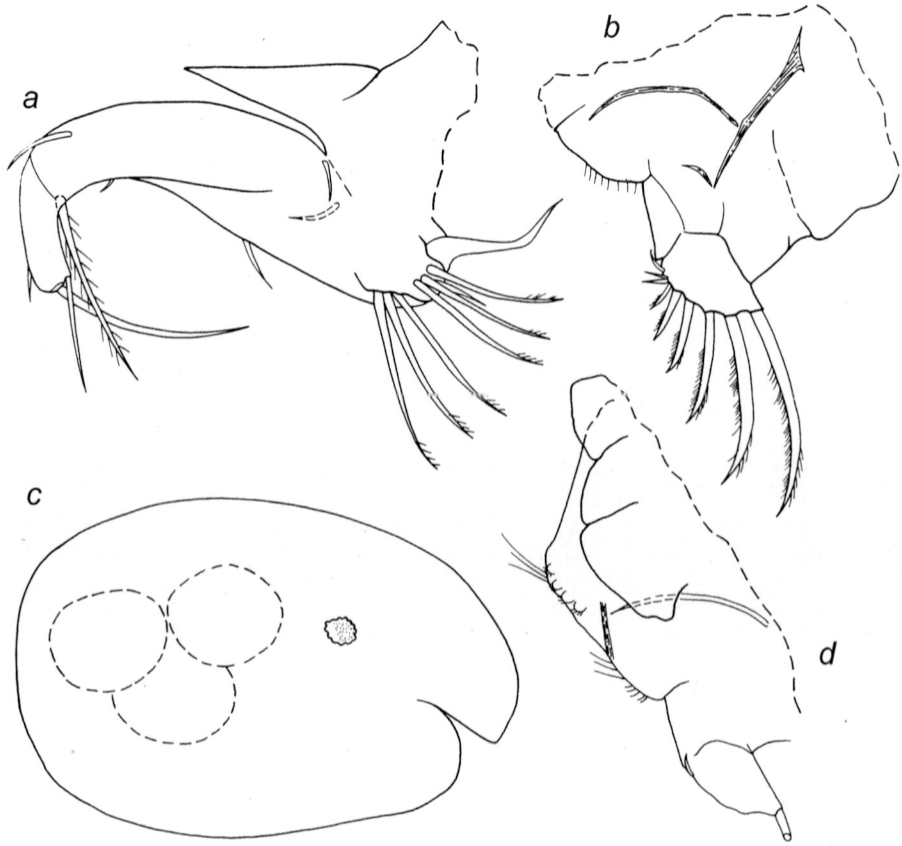

Figure 50. *Parasterope iota,* new species, holotype, BPBM-S 12866, adult female: *a,* right maxilla, mv (nabs); *b,* posterior of body from right side and right furcal lamella, lv. Paratype, USNM 1085963, adult female: *c*, complete specimen from right side, length 1.20 mm; *d*, posterior of body from right side, not all claws shown on furca.

fica bears a row of stout spines near the distal edge of articles 2 and 3. The dorsal margin of the basis of the mandible of *P. iota,* is without the small bristle present on *P. omega*. The ventral margin of the basis of the mandible proximal to the U-shaped process of *P. iota,* is without a plain bristle present on *P. zeta,* and *P. theta* (male). Articles 4 to 8 of the exopod of the 2nd antenna of *P. iota* bear basal spines absent on *P. gamma* and *P. theta* (male). The proximal dorsal corner of the protopod of the 2nd antenna of *P. iota,* is without medial rows of small spines present on *P. sigma*.

Genus *Synasterope* Kornicker

Synasterope Poulsen, 1965: 401. *Nomen nudum*; genus-group name proposed after 1930 without type species designation.

Synasterope Kornicker, 1975: 440. **Type species**. *Synasterope implumis* Poulsen, 1965, by original designation.

Composition and distribution. This genus contains numerous species. Widespread between latitudes of about 42°N and 73°S. Known depth range 1–4450 m (Kornicker, 1986: 68).

Synasterope delta, new species
Figs. 51–54

Holotype. BPBM-S 12867, adult female on slide and in alcohol.

Type Locality. Sta 10, 22 Jan 2001, Kapua Channel, Waikīkī, O'ahu, Hawaiian Islands.

Paratypes. None.

Distribution. Type locality.

Description of adult female (Figures 51–54). In lateral view carapace tapers anteriorly. Small lip present along on valve edge just ventral to incisure (Figure 51*c*).

Infold: Rostral infold with about 30 bristles (Figure 51*b*). Anteroventral infold with small bristle at inner end of incisure and about 15 additional bristles, mostly near outer edge (Figure 51*c*). Ventral infold with row of about 7 bristles. Posterior infold with about 20 flaplike bristles forming row along list near inner edge of infold; a minute bristle between few pairs of flaplike bristles (Figure 51*d*). About 15 bristles forming row just within posterior edge of valve, and 2 small indistinct pegs present in same line as row of bristles. Narrow list with anterior end at midheight of anterior of valve ventral to incisure extends along ventral infold and forms anterior edge of ridge with flaplike bristles. Valve surface with minute pores, and bristles present along valve edge.

Selvage: Ventral margin of valve with lamellar prolongation of selvage with base just proximal and medial to valve edge, and with smooth outer margin.

Central Adductor Muscle Attachments (Figure 51*a,e*): Consisting of about 14 ovoid attachments.

Carapace Size (length, height in mm): BPBM-S 12867, 1.21, 0.70.

First Antenna (Figure 51*f,g*): 1st article with indistinct lateral spines. 2nd article with long spinous distal dorsal bristle. 3rd and 4th articles separated by oblique suture. 3rd article with long dorsal margin with 5 or 6 bristles and short ventral margin with longer than usual ventral bristle. Dorsal bristles of 3rd article left limb of holotype (Figure 51*g*) differs from those of right limb (Figure 51*f*): right limb with 3 proximal bristles (middle bristle without long spines) along edge followed by paired bristles (lateral bristle of pair with proximal part parallel to edge of article) with few proximal long spines and distal short spines; left limb with 2 proximal bristles along edge (both with long spines) followed by paired bristles (lateral bristle of pair with proximal part parallel to edge of article) with few proximal long spines, and then 1 long bristle with base on lateral side of edge and with proximal part parallel to edge of article (bristle bears few long spines); article 3 of left limb also bears aberrant medial bristle (with long spines) having stout abnormal proximal part. 4th and 5th articles separated by concave suture. 4th article with long dorsal bristle with few long proximal spines; ventral margin with few spines and 2 terminal bristles (1 long, 1 unusually short). Sensory bristle of 5th article with long stem and 6 terminal filaments. 6th article fused to 7th, with long medial bristle. 7th article with claw-like a-bristle, b-bristle with 3 filaments (distal edges of bristle and distal filament with marginal spines), and long c-bristle with 5 filaments (4 proximal, 1 short subterminal). 8th article without d-bristle, bare e-bristle about same length as a-bristle, long f-bristle oriented dorsally, with 3 proximal and 1 subterminal filament, long g-bristle with 4 filaments. Total bristles 17.

Remarks. Because of the presence of an abnormal 6th bristle on the 3rd article of the left limb, it is not certain whether a normal article bears 5 dorsal bristles as on the right limb of the holotype, or 6 bristles as on the aberrant left limb.

Second Antenna: Protopodite without distomedial bristle (Figure 52*a*). Endopod with 3 articles; terminal article with very long filament (Figure 52*a*). Exopod (Figure 51*i, j*): article 1 bare; bristle of article 2 reaching past article 9, with short slender marginal spines; bristles of articles 2–8 long, with ventral spines and distal natatory hairs; 9th article elongate, with 3 bristles (ventral bristle long with ventral spines and distal natatory hairs, middle and dorsal bristles short, bare). Spines on bris-

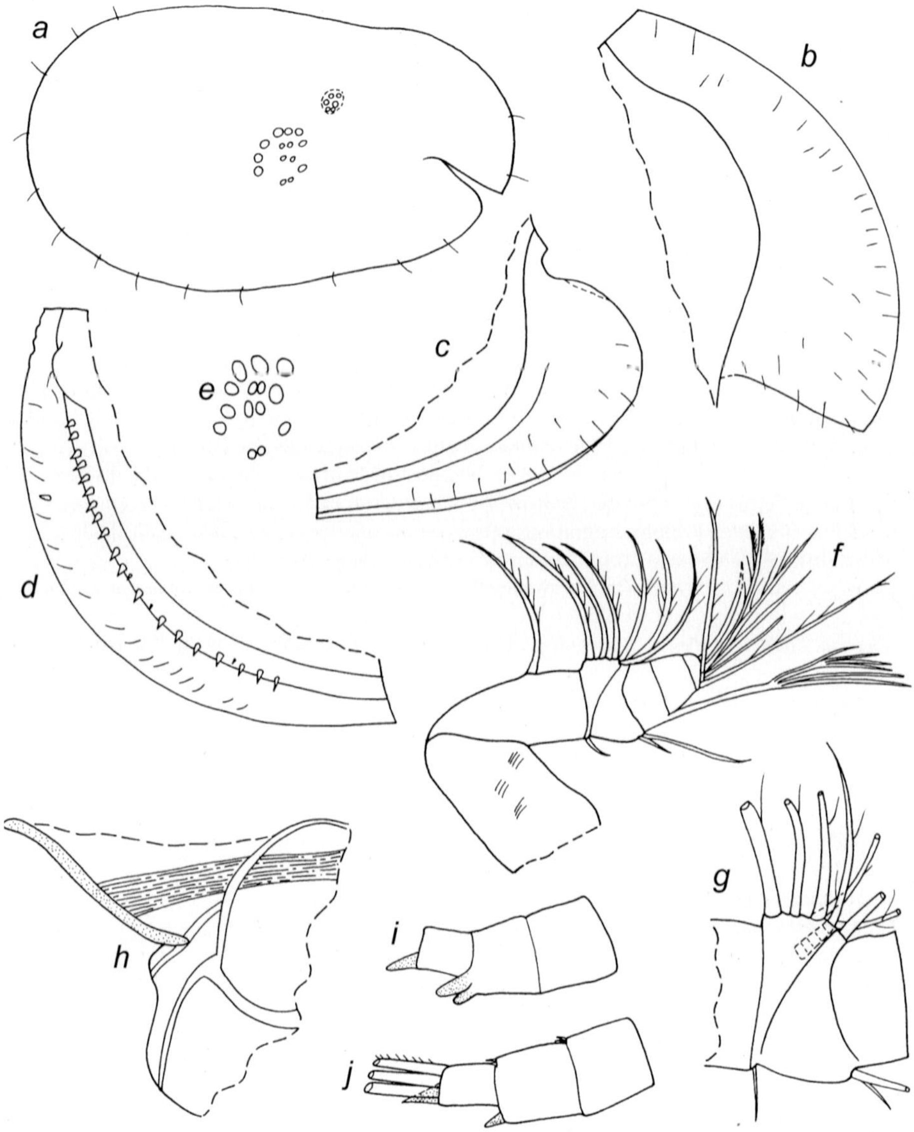

Figure 51. *Synasterope delta,* new species, holotype, BPBM-S 12867, adult female: *a,* complete specimen from right side, length 1.21 mm; *b,* rostrum left valve, iv; *c*, anteroventral corner left valve, iv; *d*, posterior left valve, iv; *e*, central adductor muscles projecting from right side of body with right valve removed; *f*, right 1st antenna, lv; *g*, part left 1st antenna, mv; *h*, pivot sclerite and proximal part of protopod left 2nd antenna, mv; *i*, distal articles exopod left 2nd antenna, lv (nabs); *j*, distal articles exopod right 2nd antenna, mv.

tles more apparent on exopod of left limb. Articles 8 and 9 with lateral basal spines (spine bifurcate on 8th article of left limb (Figure 51*i*)). Distal articles with indistinct small spines along dorsal edge (Figure 51*j*). Slightly S-shaped sclerotized pivot attached to posterior end of protopod (Figure 51*h*). Total bristles 12.

Mandible: (Figures 52*b,c,* 53*g*): Both coxa endites of BPBM-S 12867 broken off inside esoph-

Figure 52. *Synasterope delta,* new species, holotype, BPBM-S 12867, adult female: *a,* distal protopod and endopod left 2nd antenna, mv; *b,* left mandible, mv; *c,* coxa endite of mandible drawn inside esophagus (obscured); *d, e,* combs of left and right 5th limbs, lv; *f,* right 6th limb, mv; *g,* posterior tip left 6th limb, mv.

agus and obscured (Figure 53*g*); ventral branch with prolonged narrow tip (Figure 52*c*); presence or absence of small bristle usually near base of ventral branch not determined. Basis endite with 2 triaenid bristles (with 3 pairs of spines excluding terminal pair), 1 dwarf bristle, glandular peg, and 4 spinous end bristles. Basis: ventral margin with 1 triaenid bristle (with 2 or 3 pairs of marginal spines proximal to terminal pair) proximal to U-shaped process; dorsal margin with 2 long terminal bristles. Exopod reaching one half to two-thirds length of dorsal margin of 1st endopod article, hirsute, with 2 small subterminal bristles. Endopod: 1st article with 3 spinous ventral bristles (shortest with short spines, others with long spines). 2nd article: ventral margin with 3 spinous terminal bristles; dorsal margin with a-, b-, c-, and d-bristles (base of c-bristle slightly stouter than d-bristle), no lateral bristle between b- and c-bristles, 1 long lateral bristle between c- and d-bristles, 1 short bristle proximal to a-bristle, 1 short medial bristle adjacent to base of b-bristle, 6 medial cleaning bristles adjacent to c-bristle, and 1 long medial bristle adjacent to base of d-bristle. 3rd article with dorsal claw, 3 stout terminal bristles, 1 short medial bristle near ventral margin, and 1 long slender spinous bristle just ventral to base of claw. Total bristles 37.

Maxilla (Figure 53*a,b*): Endite I with 1 or 2 long bristles and 1 short bristle; endite II with 3 long bristles. Epipod narrow and protracted distally (hairs observed only on epipod of right limb). Basis: medial side with short proximal bristle near midheight, small distal bristle near ventral margin, and long spinous bristle at anterior end of row of comb bristles; lateral side with proximal backward-pointing bristle near ventral margin, and 3 dorsal bristles (1 proximal, 2 distal). Suture separating basis and endopod absent or only partly developed. Endopod: no suture separating 1st and 2nd articles; 1st article with small alpha-bristle and long beta-bristle; 2nd article with long bristle reaching well past tip of beta-bristle. Comb with 81 bristles plus proximal stout bristle with bent tip. Total bristles 97.

Fifth Limb (Figure 52*d,e*): Epipod with about 53 bristles; oblique section with about 19 bristles. Comb: surface with stout spinous exopod bristle, 1 slender bristle just ventral to base of exopod bristle, 4 bristles closer to ventral margin and just distal to base of exopod bristle, and 1 proximal and 1 distal bristle near ventral margin; ventral margin of comb with about 32 bristles; tip of comb hirsute. Total approximate bristles 111.

Sixth Limb (Figure 52*f,g*): Anterior margin with 2 endites, each with 1 bristle. Anteroventral tip of ventral margin with 2 spinous bristles (posterior of these about one-half length of other); ventral margin with 8 (right limb) or 9 (left limb) spinous bristles near midlength; posterior tip of ventral margin of left limb only with terminal bristle; medial surface with minute bristle set back from anterodorsal corner of limb. Lateral flap with hairs but no bristles. Margins of limb and medial surface hirsute. Total bristles 13–15.

Seventh Limb (Figure 53*c*): Proximal group with 6 bristles (3 on each side), with 2 to 4 or 3 to 4 bells; terminal group with 6 bristles (4 on terminal segment with 3 or 4 bells, 2 on subterminal segment with 3 bells). Terminus with opposing combs, each with about 15 narrow closely spaced teeth with indistinct marginal spines.

Furca (Figure 54*a–c*): Each lamella with 7 claws followed by 3 bristlelike claws.

Bellonci Organ (Figure 53*e*): Elongate, but obscured and exact length uncertain.

Eyes: Lateral eye with very small amount of scattered brown pigment and 6 or 7 ommatidia, about two-thirds diameter of medial eye (Figures 51*a*, 53*d,f*). Medial eye with dark brown pigment along edges and on midline (Figure 53*e,f*).

Gills (Figure 54*e*): 7 gills on each side; distal end of each gill slightly rounded.

Upper Lip (Figure 53*g*): Two rounded hirsute lobes, each with 2 small anterior spines.

Posterior of Body (Figure 54*a*): Posterodorsal corner with long spines; additional spines along posterior edge. No crescent processes present.

Y-Sclerite (Figure 54*b,c*): With pointed anterior end.

Genitalia (Figure 54*b–d*): Holotype with area with light amber color and appears to contain wormlike parasites.

Eggs (Figure 54*b,c*): Holotype with several large round ?eggs within body, each containing abundant minute cells.

Comparisons. The new species *S. delta* has only 6 or 7 ommatidia in the lateral eye. Many members of the genus are without lateral eyes, but those previously described having them, have eyes with more than 7 ommatidia. The 1st antenna does not have a lateral bristle on the 2nd article, one of the 2 ventral bristles on the 4th article is unusually small, and the stem of the sensory bristle of the 5th article is quite long, all characters useful in characterizing the species. The dorsal margin of the 3rd article of the 1st antenna may bear only 5 bristles, one less than usual, but this is not known

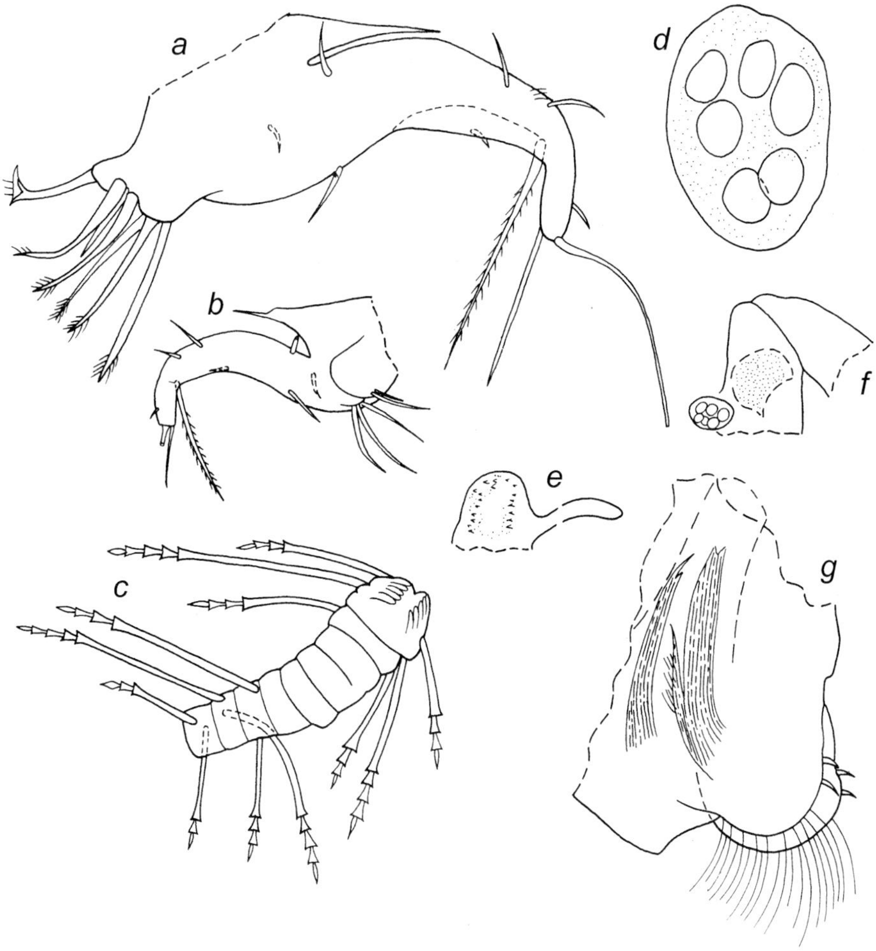

Figure 53. *Synasterope delta,* new species, holotype, BPBM-S 12867, adult female: *a,b,* right (top) and left maxillae, lv (nabs); *c,* 7th limb; *d,* left lateral eye; *e,* medial eye and Bellonci Organ; *f,* right lateral eye, medial eye (stippled), and 1st and 2nd articles of right 1st antenna; *g,* upper lip from right side showing coxa endites (striated) of mandibles inside esophagus.

with certainty because of an aberrant 6th bristle on one of the limbs of the holotype. The mandible is without a midbristle on the dorsal margin of the basis, and the basis endite bears only 1 dwarf bristle, which are additional characters useful in characterizing the species. Poulsen (1965: 410) described *S. quatrisetosa* Poulsen from an adult male collected in Apia Harbor, Samoa. The adult male is not known for *S. delta*. The length of the carapace of the male *S. quatrisetosa* is 1.71 mm, much longer than the female *S. delta* (length 1.21 mm). Also, unlike *S. delta, S. quatrisetosa* has a medial bristle on the protopod of the 2nd antenna.

Remarks concerning *Synasterope variabilis* Kornicker. Kornicker (1970: 19) referred to *Synasterope variabilis* Kornicker specimens from the Philippine Islands that have considerable differences in the number and type of bristles on the basis of the mandible (Kornicker, 1970: Table 5). The senior author now believes that the differences are too great for paratype USNM 125626 to be conspecific with the holotype (USNM 125625), and herewith refers the paratype to *Synasterope* sp.

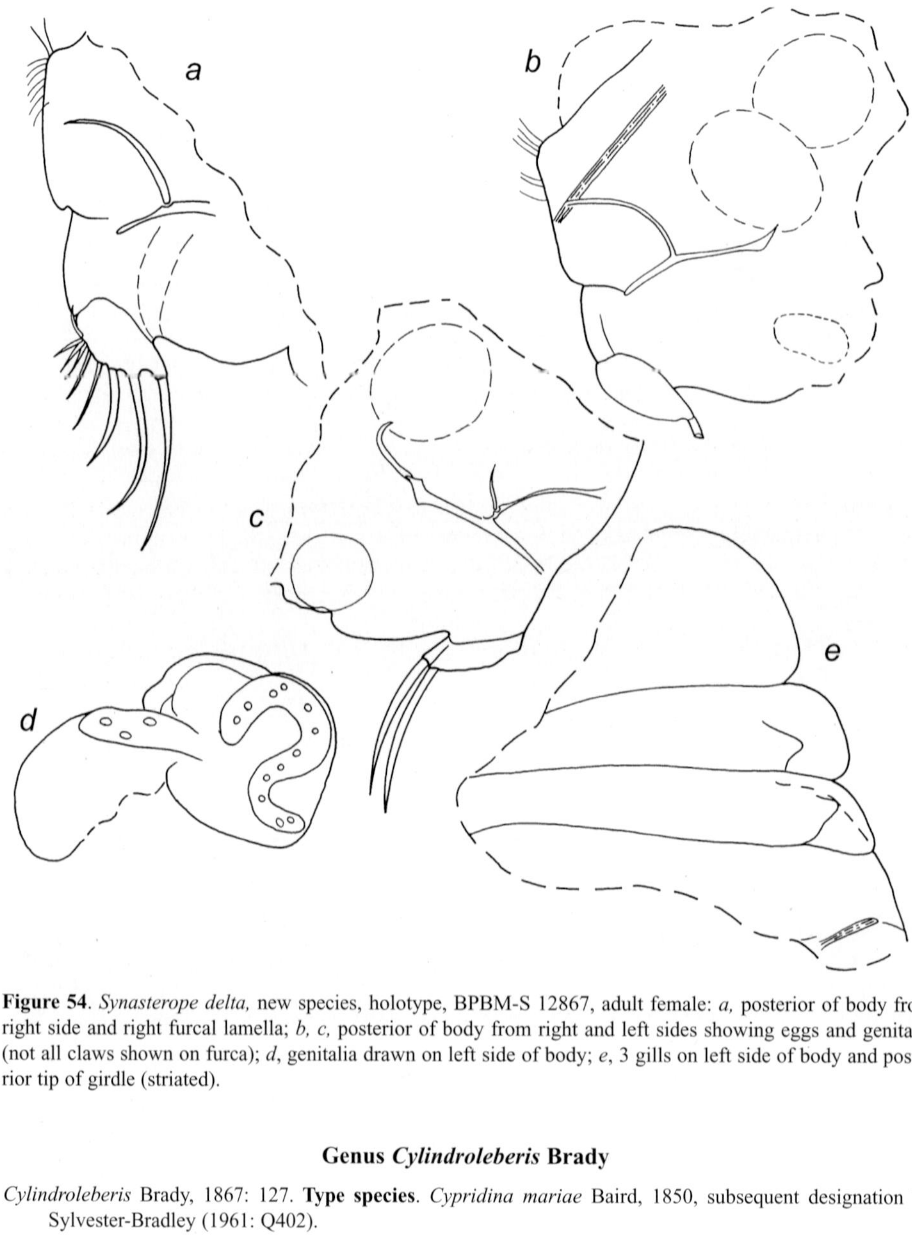

Figure 54. *Synasterope delta,* new species, holotype, BPBM-S 12867, adult female: *a,* posterior of body from right side and right furcal lamella; *b, c,* posterior of body from right and left sides showing eggs and genitalia (not all claws shown on furca); *d*, genitalia drawn on left side of body; *e*, 3 gills on left side of body and posterior tip of girdle (striated).

Genus *Cylindroleberis* Brady

Cylindroleberis Brady, 1867: 127. **Type species**. *Cypridina mariae* Baird, 1850, subsequent designation by Sylvester-Bradley (1961: Q402).

Composition and distribution. Including the new species described herein the genus includes 12 species and one subspecies (Kornicker, 1991: 110; Kornicker & Thomassin, 1998: 111). Circumglobal between latitudes of about 58°N and 26°S. Generally, part of the benthos in shallow water (Kornicker, 1991: 110; Kornicker & Thomassin, 1998: 111).

Cylindroleberis kappa, new species

Figs. 55–67

Holotype. BPBM-S 12836, ovigerous female on 3 slides and in alcohol.

Type Locality. Sta 11, Kāne'ohe Bay, O'ahu, Hawaiian Islands.

Paratypes. Sta 2: USNM 1085964, ovigerous female on 2 slides and in alcohol; USNM 1085965, instar III female on 2 slides; BPBM-S 12820, 6 specimens (5 ovigerous female, 1 adult female) in alcohol. Sta 11: USNM 1085966, ovigerous female on 2 slides and in alcohol; BPBM-S 12837, 2 ovigerous females in alcohol. Sta 12: USNM 1085967, 1 adult male on 3 slides and in alcohol. Sta 17: BPBM-S 12825, ovigerous female in alcohol; USNM 1085968, instar VI male on 2 slides and in alcohol; BPBM-S 12826, instar VI male in alcohol; USNM 1085969, instar V male on 2 slides and in alcohol; BPBM-S 12827, instar III female in alcohol. USNM 1085970, instar IV male on 2 slides and in alcohol; BPBM-S 12828, 39 specimens in alcohol.

Distribution. Kāne'ohe Bay, Sta 2, 11, 12, 17.

Description of adult female (Figures 55–59). Carapace elongate with parallel ventral and dorsal margins (Figures 55*a, 58a,* 59*a*).

Infold: Infold of rostrum with 21 to 30 long and 11 small bristles (Figure 55*b*). Anteroventral infold with about 20 long bristles and 8 small bristles (Figure 55*c*). Ventral infold with about 18 bristles forming row with posterior bristle near anterior flaplike bristle of posterior infold. Posterior infold with broad band along inner end of infold with about 13 flaplike bristles and about 10 minute bristles (not more than 1 bristle between each pair of flaplike bristles); 5 knoblike processes forming row between broad band and outer edge of valve (Figure 55*d*).

Selvage: Narrow infold with smooth outer edge present along ventral margin. None observed in vicinity of incisure. Posterior infold near dorsal margin with short hirsute section on right valve (left valve obscured).

Central Adductor Muscle Attachments (Figure 55*a*): Comprising about 12 ovoid attachments.

Carapace Size (length, height in mm): BPBM-S 12836, 1.37, 0.58. USNM 1085966, 1.22, 0.54. USNM 1085964, 1.27, 0.62. BPBM-S 12837a, 1.22, 0.57. BPBM-S 12837b, 1.28, 0.56.

First Antenna (Figure 56*a*): 1st article bare. 2nd article with 1 long spinous distal dorsal bristle and 1 short distal medial bristle. 3rd and 4th articles separated by oblique suture better developed on lateral side than on medial side. 3rd article with short ventral margin with small bristle and long dorsal margin with 6 bristles (5 bristles along edge (middle bristle bare, others with long spines) and 1 lateral bristle (with short spines) adjacent to 4th edge bristle). Suture on lateral side separating 4th and 5th articles linear and distinct; suture on medial side represented by short line near ventral margin. 4th article with short dorsal margin with 1 long bristle (bare or with indistinct short spines) and long ventral margin with 2 bare terminal bristles (1 long, 1 short). Sensory bristle of 5th article with 1 short slender proximal filament and 6 long stouter terminal filaments, latter with minute pointed terminal process. 6th article with long bare terminal medial bristle. 7th article: a-bristle clawlike with minute lateral teeth along concave margin; b-bristle with 1 short and 3 long marginal filaments; c-bristle long with 5 marginal filaments (4 proximal long, 1 subterminal short). 8th article: d-bristle absent; e-bristle bare, longer than a-bristle and shorter than b-bristle; f-bristle bent dorsally, with 4 marginal filaments; g-bristle long with 4 marginal filaments. Total bristles 20.

Second Antenna: Protopod bare (Figure 56*b*). Endopod with 2 articles (Figures 56*b,* 58*b*); 2nd article with terminal filament. Exopod: bristle of 2nd article reaching past 9th article, with marginal spines; bristles of articles 3 to 5 with slender ventral spines and natatory hairs; bristles of articles 6 to 8 with natatory hairs and few indistinct slender ventral spines; 9th article with 4 bristles (dorsal bristle short bare, adjacent bristle of medium length with few natatory hairs, adjacent bristle long and with natatory hairs; ventral bristle long with slender ventral spines and natatory hairs); articles 3 to

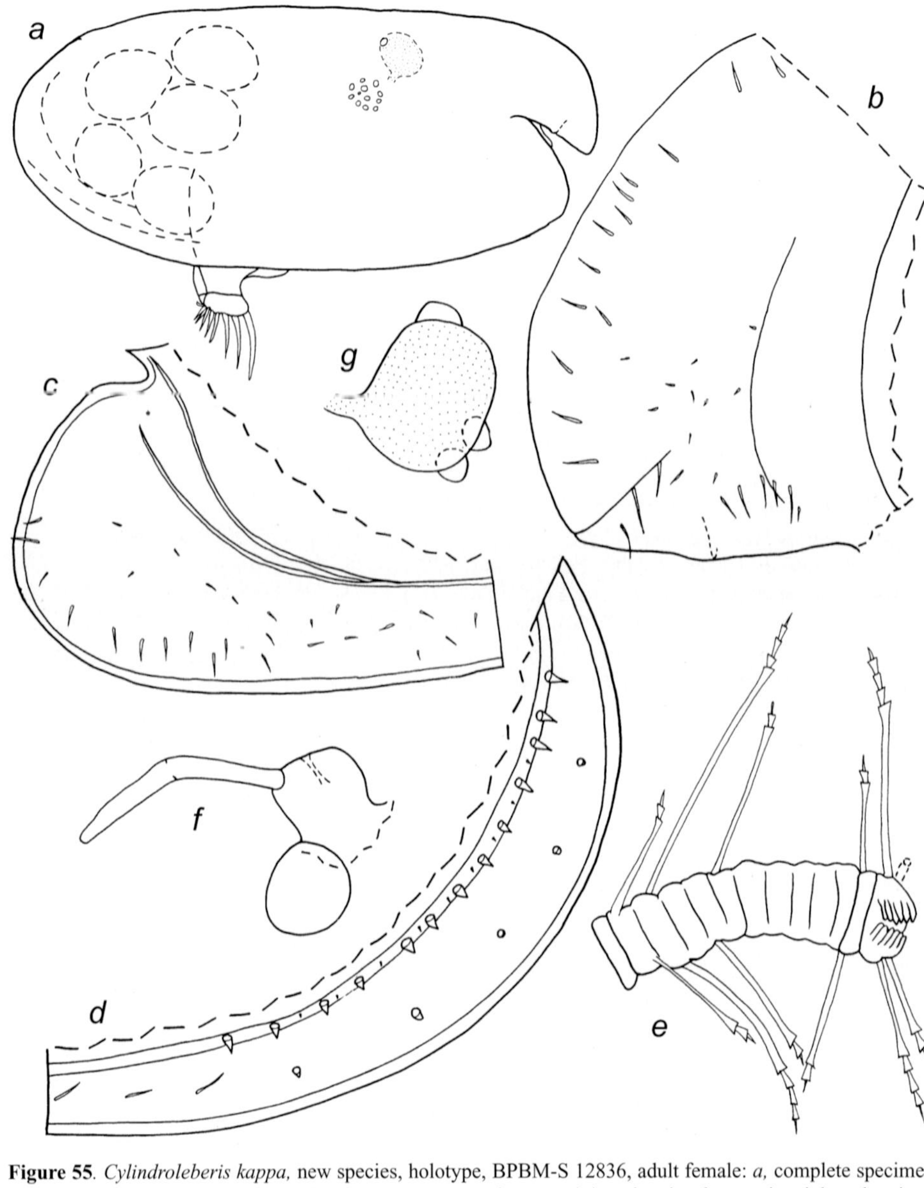

Figure 55. *Cylindroleberis kappa*, new species, holotype, BPBM-S 12836, adult female: *a*, complete specimen, length 1.37 mm; *b*, rostrum, right valve, iv; *c*, anteroventral corner right valve, iv; *d*, posterior right valve, iv; *e*, 7th limb; *f*, outline left lateral eye, medial eye, and Bellonci Organ; *g*, right lateral eye.

8 with basal spines (Figure 56*c*); spine of 8th article about two-thirds length of 9th article; 9th article with bifurcate lateral spine about three-fourths length of article; articles 3 to 8 with minute terminal spine on dorsal edge. Total bristles 13.

Mandible (Figures 57*a,c,g*, 58*c*): Coxa endite: small medial bristle near base of ventral branch; ventral branch with 4 or 5 rows of ventral spines and trident tip. Dorsal branch with 3 rounded ventral nodes and short main spine; tip with terminal bristle and longer dorsal bristle (Figure 58*c*). Basis endite with knob adjacent to 2 dwarf bristles, 2 or 3 triaenid bristles (2 posterior to knob, 0 or 1 ante-

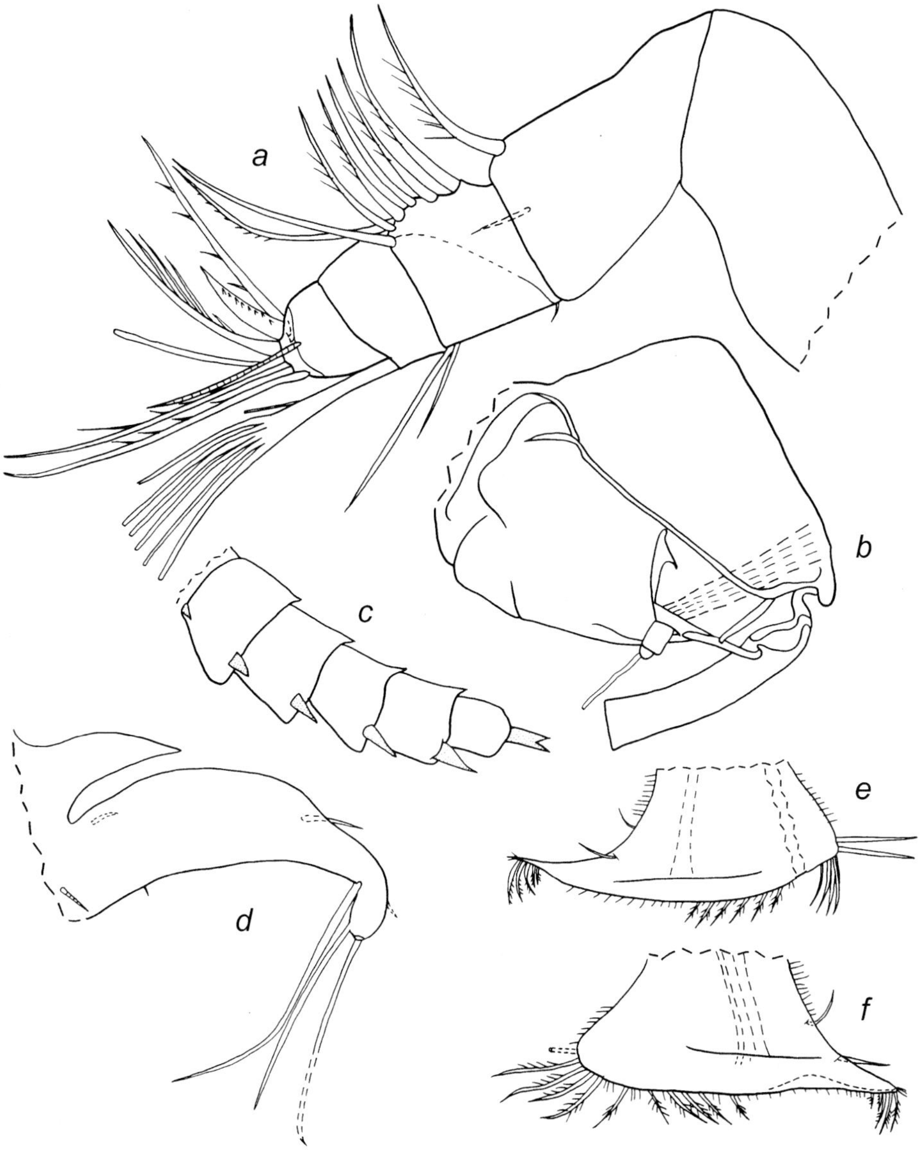

Figure 56. *Cylindroleberis kappa,* new species, holotype, BPBM-S 12836, adult female: *a,* right 1st antenna, mv; *b,* left 2nd antenna, mv; *c,* distal exopod right 2nd antenna (nabs); *d,* right maxilla, lv (nabs); *e,f,* left and right 6th limbs, lv.

rior to knob) with 2 to 4 spines on each side proximal to terminal pair (BPBM-S 12836 with 2 triaenid bristles, USNM 1085966 and 1085964 with 3), and 5 spinous end bristles (Figure 57*a,b*). Basis with 1 small bare ventral bristle; dorsal margin with 1 small bare bristle at midlength and 2 long terminal bristles with few indistinct small marginal spines. Exopod small, about one-fourth length of dorsal margin of 1st endopod article, with 2 small terminal bristles (Figure 57*a,c*). Endopod: 1st article with 3 long ventral bristles (2 with long spines, 1 with short spines). 2nd article: ventral margin with 3 long terminal bristles with short marginal spines; dorsal margin with stout a-, b-, c-, and d-

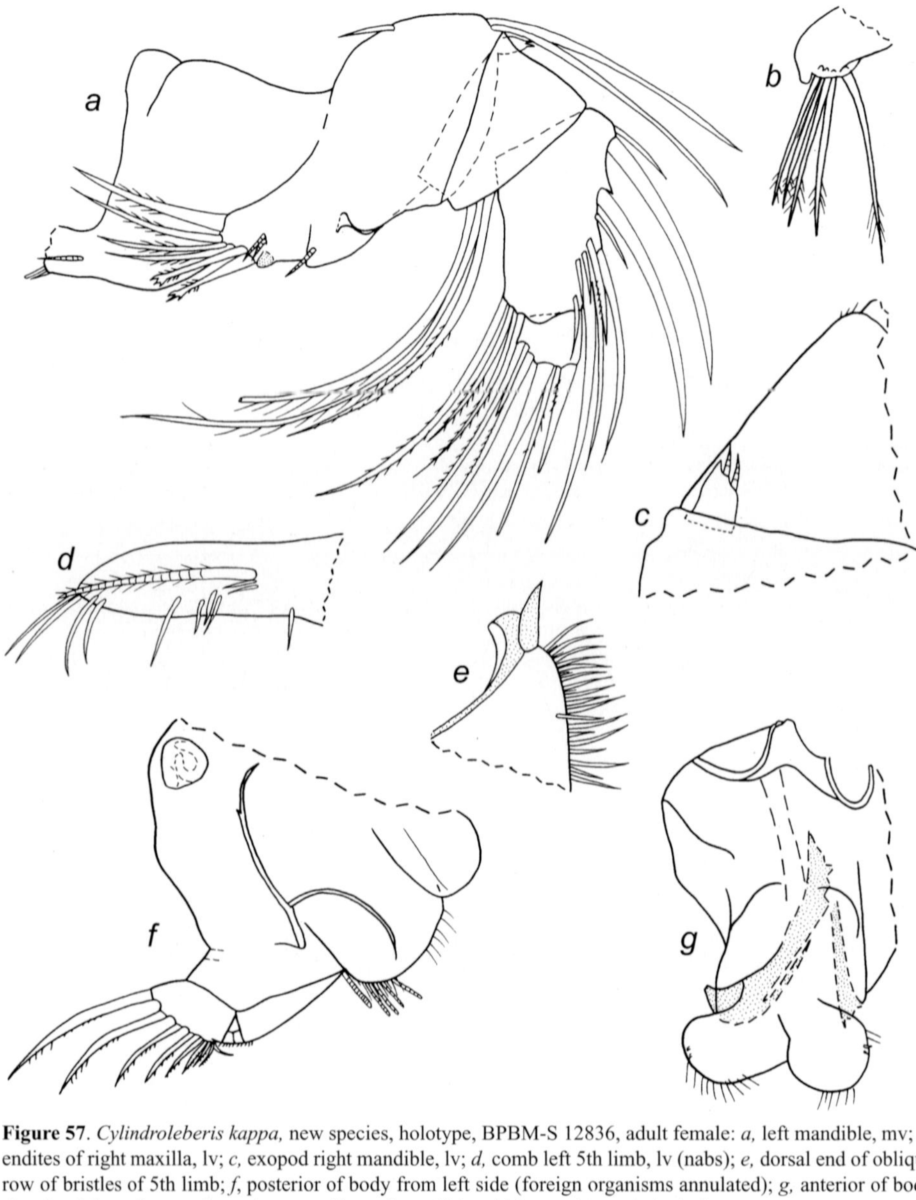

Figure 57. *Cylindroleberis kappa,* new species, holotype, BPBM-S 12836, adult female: *a,* left mandible, mv; *b,* endites of right maxilla, lv; *c,* exopod right mandible, lv; *d,* comb left 5th limb, lv (nabs); *e,* dorsal end of oblique row of bristles of 5th limb; *f,* posterior of body from left side (foreign organisms annulated); *g,* anterior of body showing lobes of upper lip and coxa endites of mandible within esophagus (stippled).

bristles (base of c-bristle only slightly broader than bases of b- and d-bristles; d-bristle with short spines, others bare), 1 small bristle proximal to a-bristle, 1 or 2 short medial bristles (with short spines) between b- and c-bristles, 3 or 4 short medial bristles (with marginal spines) forming row adjacent to c-bristle, 1 short medial bristle adjacent to d-bristle, 1 long lateral bristle between c- and d-bristles; no long lateral bristle between b- and c-bristles. 3rd article with stout straight dorsal claw with few proximal spines along ventral edge, 3 long terminal bristles, 1 short medial bristle near ventral margin, and 1 long slender lateral bristle just ventral to stout dorsal claw. Total bristles 40 to 43.

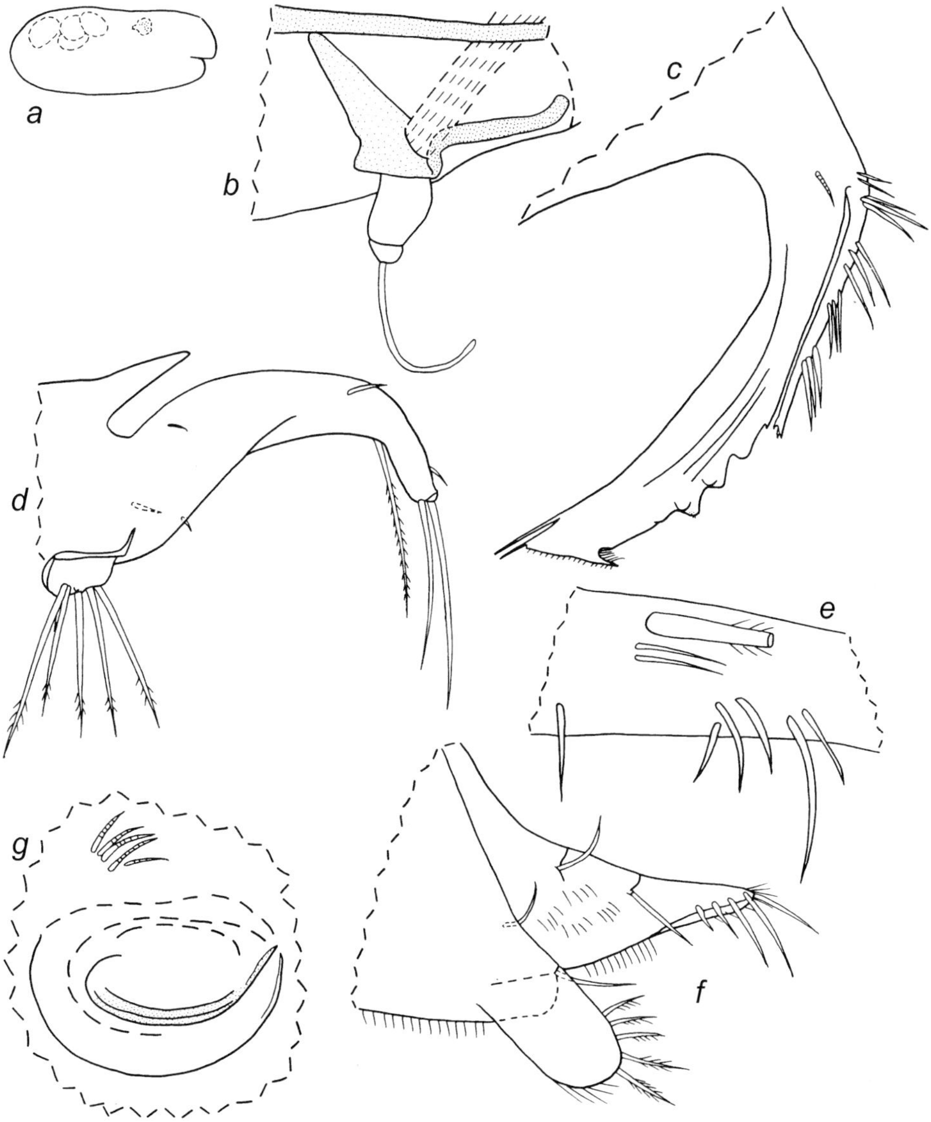

Figure 58. *Cylindroleberis kappa,* new species, paratype, USNM 1085966, adult female: *a,* complete specimen, length 1.22 mm; *b,* endopod left 2nd antenna, mv; *c,* coxa endite, left mandible, mv; *d,* left maxilla, mv (nabs); *e,* part comb of right 5th limb (nabs), lv; *f,* anterior left and right 6th limbs, under cover slip; *g,* genitalia and bristles left side of body.

Maxilla (Figures 56*d,* 57*b,* 58*d*): Endite I with 2 long and 1 short bristle; endite II with 3 long bristles. Epipod short triangular. Basis with short proximal lateral bristle near midwidth, 1 short proximal medial bristle near dorsal margin, 1 short distal dorsal bristle, 1 small proximal ventral bristle with base on lateral side, and 1 long ventral terminal bristle. Endopod: 1st article with small alpha-bristle and 1 long beta-bristle; 2nd article with long terminal bristle slightly longer than beta-bristle. Comb with 62 bristles plus proximal end bristle with bent tip. Total bristles about 77.

Fifth Limb (Figures 57*d,e,* 58*e*): Epipod with 51 hirsute bristles; oblique section with about 17

a

b

c

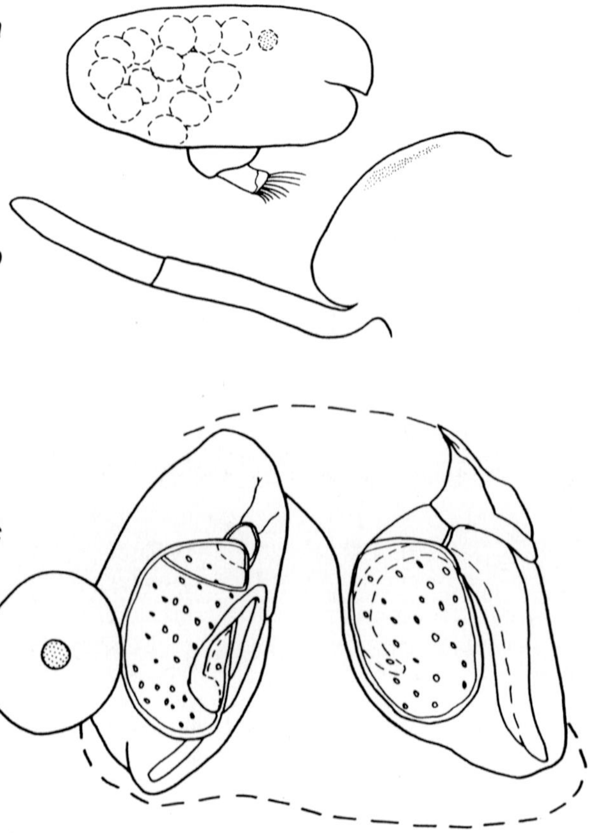

Figure 59. *Cylindroleberis kappa,* new species, paratype, USNM 1085964, adult female: *a,* complete specimen, length 1.27 mm; *b,* medial eye and Bellonci Organ; *c,* ventral view of left and right genitalia with attached ?egg.

spinous bristles along hirsute edge and 1 near midheight set back from edge (Figure 57*e*). Comb (Figures 57*d,* 58*e*): with stout spinous exopod bristle, 2 small slender bristles just ventral to base of exopod bristle, and 7 or 8 bristles near ventral margin (1 proximal, 4 or 5 near midlength, and 2 distal); ventral margin of comb with about 25 spinous bristles. Total bristles about 104.

Sixth Limb (Figures 56*e,f,* 58*f*): Anterior margin with 2 endites, each with 1 bristle. Anterior tip of ventral margin with 5 spinous bristles (3 long, 2 short); lateral flap hirsute but without bristles. Ventral and posteroventral margin with 11 or 12 spinous bristles, those on posterior corner longer. Anterior, ventral and posterior margins hirsute. Total bristles about 18.

Seventh Limb (Figure 55*e*): 6 bristles in proximal group (with 2 to 4 bells); 6 bristles in distal group (with 2 to 4 bells). Terminus with opposing combs, each with about 12 or 13 spinous teeth. Total bristles 11 or 12.

Furca (Figure 55*a,* 57*f,* 59*a*): Each lamella with 7 claws followed by 2 bristlelike claws (last bristle pointing backward). Total claws and bristles 9.

Bellonci Organ (Figures 55*f,* 59*b*): Elongate with rounded tip and indistinct suture near midlength.

Eyes: Lateral eye with many ommatidia (about 20) obscured by black pigment (Figures 55*a,f,g,* 58*a,* 59*a*). Medial eye about same size as lateral eye, with small amount of colored area, bare (Figures 55*f,* 59*b*).

Upper Lip (Figure 57*g*): 2 hirsute lobes with 2 minute anterior spines. (Figure 57*g* is anterior view of body with upper lips flattened (anterior ends of lips located on each side), and shows coxa endites of mandibles within esophagus.)

Anterior of Body (Figure 57*g*): Bare.

Posterior of Body (Figure 57*f*): Spinous. BPBM-S 12836 with numerous foreign filamentlike annulated organisms attached to posterior margin (not all shown in Figure 57*f*).

Y-Sclerite (Figure 57*f*): With elongate curved anterior end and small ventral branch at about one-third length from anterior end.

Genitalia (Figure 57*f*, 58*g*, 59*c*): Amber colored area anterior to furca. BPBM-S 12836 with internal tubular structure (Figure 57*f*). USNM 1085966 with row of 5 minute bristles just anterior and dorsal to genital area (Figure 58*g*). USNM 1085964 with oval spermatophore containing minute round sperm in each genital organ (Figure 59*c*).

Gills: 6 or 7 on each side, well developed with convex distal ends (Figure 57*f*).

Eggs (Figures 55*a*, 58*a*, 59*a*): BPBM-S 12836 with 6 eggs of similar size in marsupium; length and height of 1 egg (mm): 0.220, 0.156. USNM 1085966 with 4 eggs in marsupium: length and height of 2 eggs (mm): 0.190, 0.142; 0.181, 0.171. USNM 1085964 with 21 eggs: length and height of 1 egg (mm): 0.222, 0.170. BPBM-S 12837 with about 10 eggs. BPBM 12837, specimen b with 4 eggs. BPBM-S 12825 with about 15 eggs in marsupium. Some specimens with unextruded small eggs in addition to large eggs in marsupium.

Gut Content: Light amber color.

Foreign Organisms: BPBM-S 12836 with many segmented filaments attached to appendages and body (Figure 57*f*).

Description of adult male (Figures 60–62). Except for slightly greater overhang of rostrum, carapace similar in outline to that of adult female, and with vertical row of bristles near dorsal end (Figure 60*a*).

Infold: In general, similar to that of adult female.

Selvage: Lamellar prolongation along ventral margin as on adult female.

Central Adductor Muscle Attachments (Figure 62*c*): With 13 attachments.

Carapace Size (length, height in mm): USNM 1085967, 1.50, 0.71.

Body length (Figure 60*a*): Including extended c- and g-bristles of 1st antenna: 3.48 mm.

First Antenna (Figures 60*a–d*, 62*b*): 1st article bare. 2nd article with long spinous dorsal bristle and short distal lateral bristle. 3rd article with small bristle on short ventral margin and 6 bristles on long dorsal margin. 4th article with 2 ventral bristles and 1 dorsal bristle. Sensory bristle of 5th article stout with numerous hairs. 6th article with medial bristle near dorsal margin. 7th article: a-bristle clawlike on small projection; b-bristle about 3 times length of a-bristle, with 5 marginal filaments; c-bristle very long with 28 short, slender, bare filaments. 8th article: d-bristle absent; e-bristle shorter than b-bristle, bare with blunt tip; f-bristle very long with 27 short, slender, bare marginal filaments; g-bristle longer than b-bristle, with about 8 short marginal filaments. Dorsal margin of 6th article slightly undulate (Figure 60*c*). 7th article with sclerotized lateral triangular terminal process (Figure 60*c*). Total bristles 25.

Second Antenna (Figures 60*e*, 61*a,b*, 62*c*): Protopodite bare; sclerites at base of endopod differ from those of female. Endopod with 3 articles: article 1 bare; article 2 elongate with 3 short distal bristles; article 3 reflexed, with 1 distal filament. Exopod with 9 articles: article 1 bare; article 2 elongate (3 times length of article 3), with natatory bristle reaching past 9th article; articles 3 to 8 with long natatory bristle; article 9 with 4 bristles (3 long with natatory hairs, 1 short, dorsal, with marginal hairs). Articles 2 to 8 with long terminal hairs at dorsal corner; articles 3 to 9 with small terminal spine at ventral corner; some distal articles with small basal spines (obscured). Total bristles 16.

Mandible (Figures 60*f*, 61*c,d*): Coxa endite similar to that of adult female (Figure 60*f*). Basis endite with 3 or 4 end bristles with indistinct short marginal hairs, 2 short bare bristles, indistinct glandular peg adjacent to 2 bare dwarf bristles (Figure 61*c,d*). Basis: dorsal margin with 1 short backward pointing midbristle and 2 long terminal bristles; ventral margin with 1 small pointed prox-

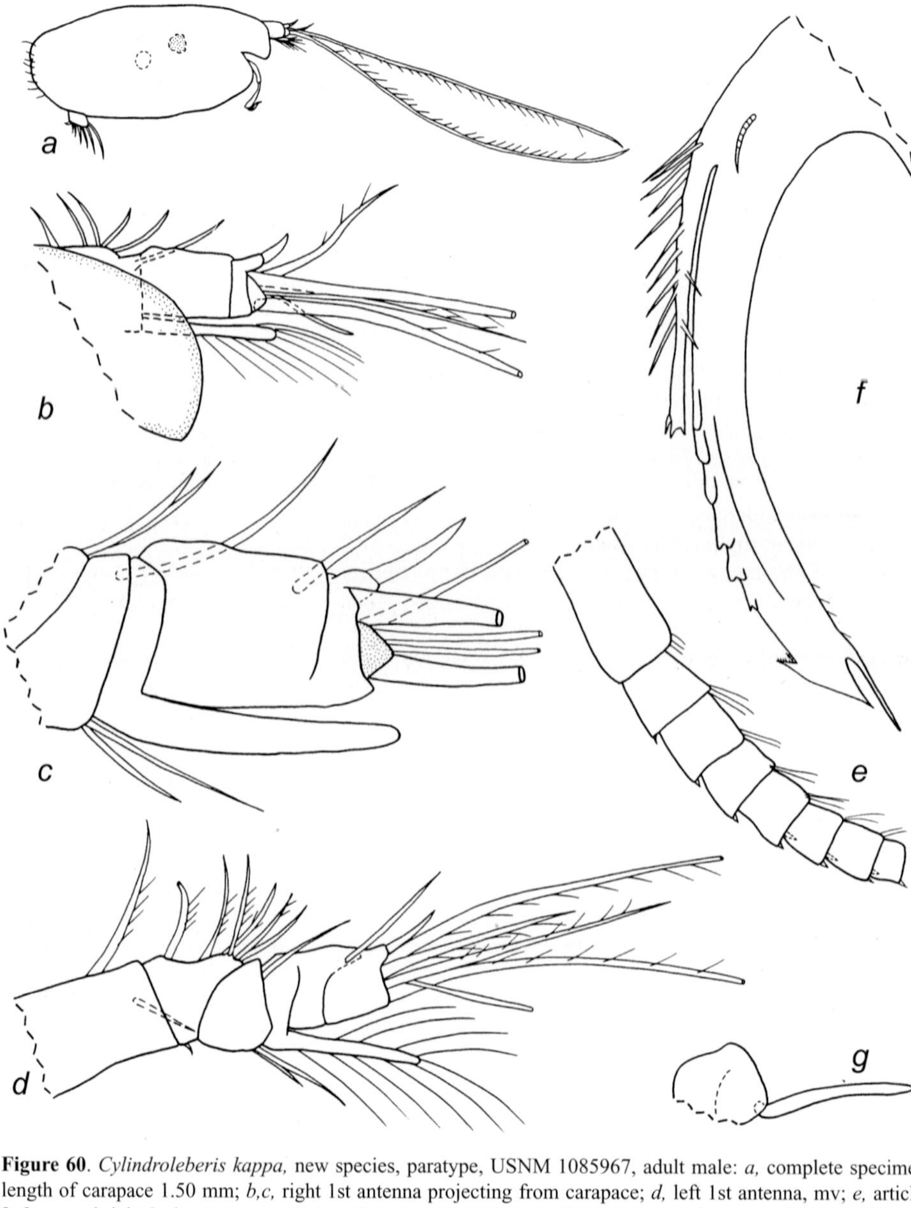

Figure 60. *Cylindroleberis kappa,* new species, paratype, USNM 1085967, adult male: *a,* complete specimen, length of carapace 1.50 mm; *b,c,* right 1st antenna projecting from carapace; *d,* left 1st antenna, mv; *e,* articles 2–9 exopod right 2nd antenna, mv (nabs); *f,* coxa endite right mandible, mv; *g,* medial eye and Bellonci Organ.

imal bristle very close to base of endite of basis. Exopod similar to that of adult female, about one-fourth length of dorsal margin of 1st article of endopod, hirsute, with 2 short terminal bristles. Endopod: 1st article with 3 ventral bristles (2 long with stout spines along anterior edge, 1 about one-half length of others and with slender proximal spines along both margins). 2nd article: ventral margin with 3 terminal bristles; dorsal margin with stout a-, b-, c-, and d-bristles (base of c-bristle about same width as bases of a- and b-bristles, slightly broader than base of d-bristle); 3 slender bristles (less than one-half length of a-bristle) on dorsal edge proximal to a-bristle; lateral side with long lat-

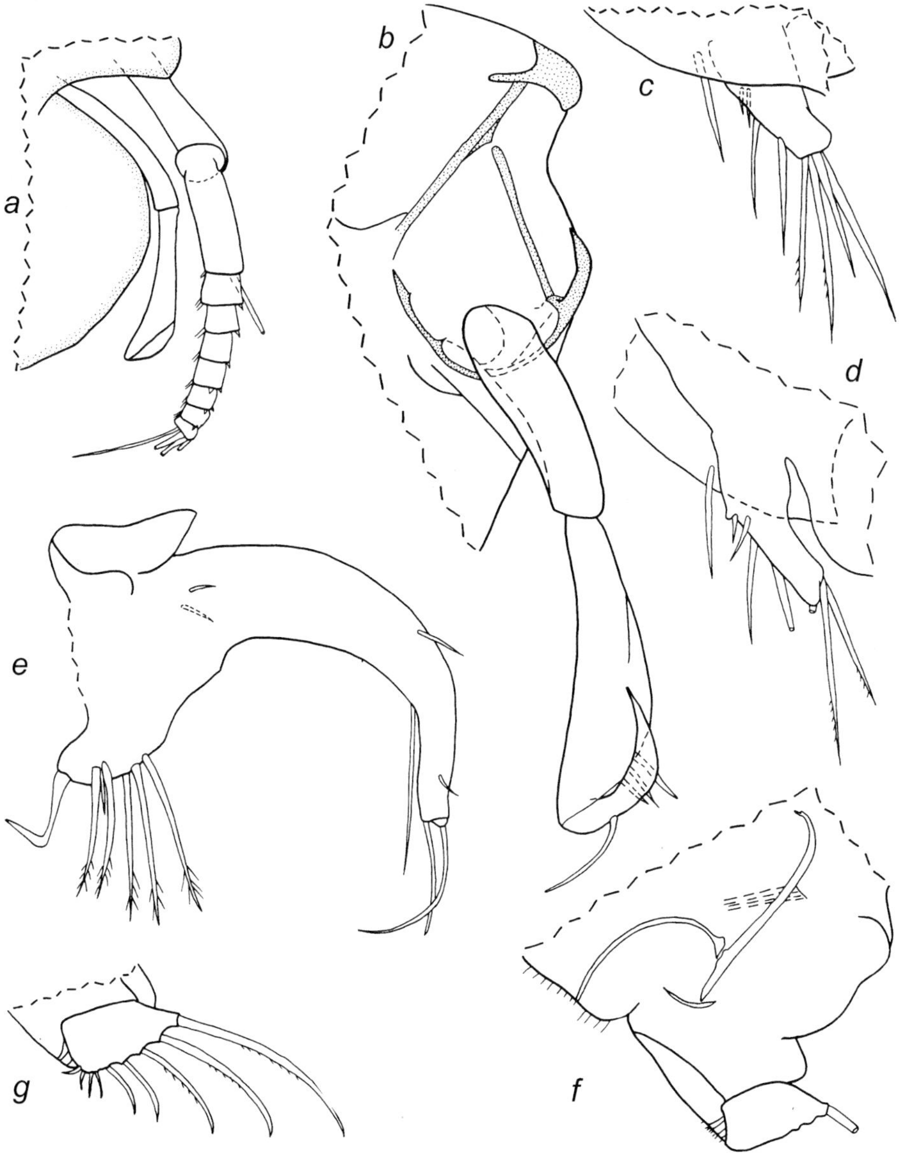

Figure 61. *Cylindroleberis kappa,* new species, paratype, USNM 1085967, adult male: *a,* right 2nd antenna projecting from carapace, lv (nabs); *b,* distal protopod and endopod left 2nd antenna, mv; *c,* basis endite left mandible, lv; *d,* basis endite right mandible, mv; *e,* left maxilla, mv; *f,* posterior of body from left side, not all claws of furca shown; *g,* right lamella of furca, lv.

eral bristle between c- and d-bristles; medial side with 5 spinous bristles forming 1 or 2 oblique rows between b- and d-bristles and 1 short spinous bristle adjacent to d-bristle. 3rd article with stout straight dorsal claw, 3 long end bristles, 1 short medial bristle near ventral edge, and 1 long lateral bristle just ventral to base of dorsal claw. Total bristles 42 or 43.

Maxilla (Figure 61*e*): Epipod short (folded in Figure 61*e*). Endite I with 3 bristles (2 long, 1 short); endite II with 3 long bristles. Basis with 1 proximal lateral bristle, 1 proximal medial bristle,

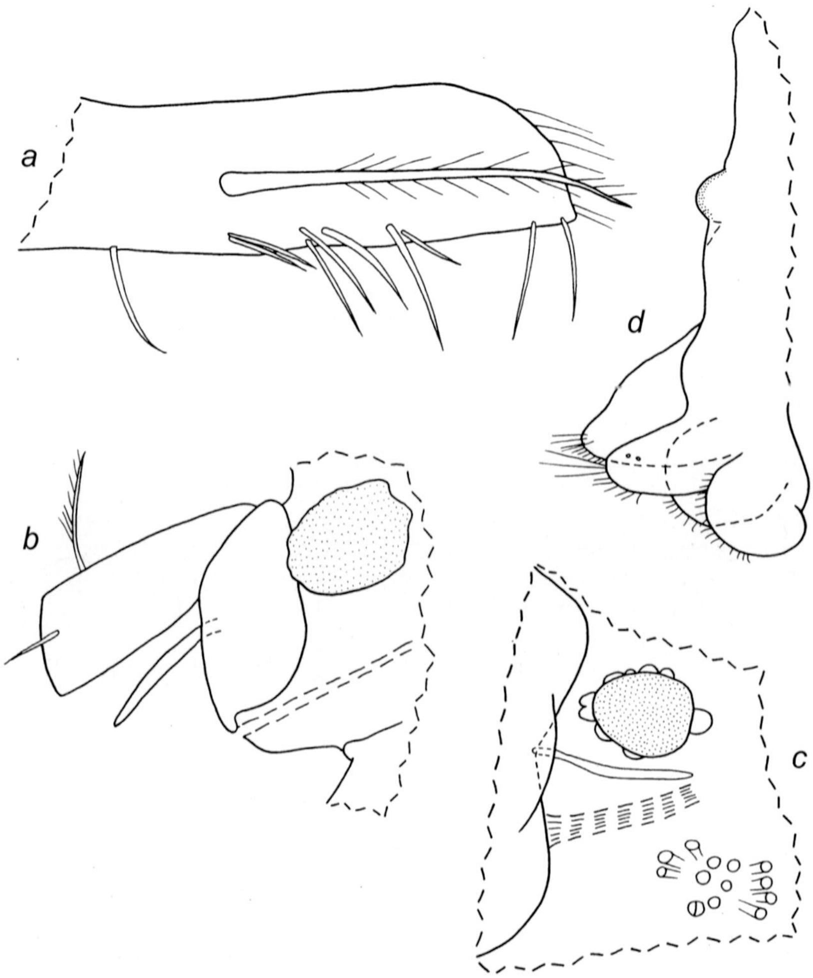

Figure 62. *Cylindroleberis kappa,* new species, paratype, USNM 1085967, adult male: *a,* comb of right 5th limb, lv (nabs); *b,* anterior of body from left side showing outline of lateral eye, part of 1st antenna, and Bellonci Organ; *c,* anterior of body from left side showing posterior end protopod 2nd antenna and attached pivot sclerite, lateral eye, and central adductor muscles; *d,* anterior of body from left side showing rounded node at midheight and upper and lower lips.

1 distal dorsal bristle, and 1 long terminal ventral bristle. Endopod: 1st article with short alpha-bristle and long beta-bristle. 2nd article with terminal bristle longer than beta-bristle. Comb with 60 bristles plus stout proximal bristle with bent tip. Total bristles 74.

Fifth Limb (Figure 62*a*): Comb with stout spinous exopod bristle reaching past tip of comb; 2 slender bristles ventral to base of exopod bristle; 5 bristles close to base of comb near midlength; 1 proximal and 2 distal bristles near base of comb. Epipod with about 47 bristles; oblique section with 20 bristles; ventral margin of comb with 31 bristles. Approximate total bristles 109.

Sixth Limb (Figure 60*a*): Shape similar to that of adult female. Anterior margin with upper and lower endite bristle. Anterior corner of ventral margin with 5 spinous bristles; lateral flap hirsute but without bristles. Ventral margin with 15 spinous bristles; bristles longer along posteroventral curvature; anterior bristle located just anterior to midwidth. Anterior, ventral, and posterior edges hirsute. Total bristles 22.

Seventh Limb: Similar to that of adult female. Total bristles 12.

Furca (Figures 60*a*, 61*f,g*): Each lamella with 8 claws followed by 1 bristlelike claw. Total claws and bristles 9.

Belloncі Organ (Figures 60g, 62*b*): Similar to that of female.

Eyes: Lateral eye large with many ommatidia obscured by black pigment except along edge (Figures 60*a*, 62*b,c*); eye larger than that of adult female. Medial eye smaller than lateral eye, bare, with small amount of pigment near dorsal edge (Figure 60*g*).

Lips (Figure 62*d*): Upper and lower lips hirsute. Spines not observed.

Anterior of Body (Figure 62*d*): With small rounded node at midheight.

Posterior of Body (Figure 61*f*): Similar to that of adult female.

Y-Sclerite (Figure 61*f*): Similar to that of adult female.

Copulatory organ: Not identified with certainty, small and may have a few indistinct bristles.

Gills: Similar to those of adult female.

Gut Content: Gut with light amber color.

Description of instar VI male (Figure 63). Carapace elongate with parallel dorsal and ventral margins and sloping posterodorsal margin (Figure 63*a*).

Infold: In general similar to that of adult female; posterior infold with small processes.

Selvage: Lamellar prolongation with smooth outer edge present along ventral margin.

Central Adductor Muscle Attachments (Figure 63*b*): With about 12 ovoid attachments.

Carapace Size (length, height in mm): USNM 1085968, 1.33, 0.61. BPBM-S 12826, 1.31, 0.66.

First Antenna (Figure 63*c,h*): Except for short part of indistinct suture near ventral margin, 3rd and 4th articles fused. Article 8: f-bristle oriented posteriorly, with about 7 minute dorsal spines followed by 3 short filaments, 3 longer filaments, and then 3 still longer filaments (not all shown). Limb otherwise similar to that of adult female. Total bristles 20.

Second Antenna (Figure 63*d,e*): Protopodite bare; sclerites in vicinity of endopod similar to those of adult female. Endopod with 3 articles: 1st and 2nd articles elongate; article 2 with 2 minute spinelike distal bristles; 3rd article with pointed tip and proximal filament reaching past end of endopod (Figure 63*d*); suture absent between articles 2 and 3. Exopod with 9 articles: article 1 bare; bristle of article 2 reaching 9th article, with ventral spines; bristle of article 3 long with proximal ventral spines and natatory hairs; bristles of articles 4–8 long with natatory hairs; 9th article with 4 bristles (dorsal bristle very short, bare; ventral bristle very long with natatory hairs; bristle dorsal to ventral bristle shorter than ventral bristle, with natatory hairs; bristle ventral to dorsal bristle fairly short, with indistinct natatory hairs). Articles 4 to 8 with pointed basal spines and spines forming row along distal margin; article 9 with pointed lateral spine (Figure 63*e*). Total bristles 15.

Mandible (Figure 63*h*): Coxa endite similar to that of adult female. Basis endite with 3 triaenid bristles (2 distal to process, 1 proximal to process), small projecting process adjacent to 2 dwarf bristles, and 4 spinous end bristles. Basis with 1 short dorsal pointing midbristle and 2 terminal dorsal bristles. Exopod similar to that of adult female. Endopod: 1st article with 3 ventral bristles (2 with long spines, 1 shorter and with short spines). 2nd article: dorsal margin with stout a-, b-, c-, and d-bristles, and 1 short bristle proximal to a-bristle; medial side with 1 short spinous bristle between b- and c-bristles, 4 spinous bristles forming row between c- and d-bristles, and 1 short bristle distal to d-bristle; lateral side with 1 long bristle between c- and d-bristles; ventral margin with 3 terminal bristles. 2nd article similar to that of adult male. Total bristles 41.

Maxilla (Figure 63*f*): Similar to that of adult female except comb with 59 ventral bristles plus proximal bent bristle. Total bristles 74.

Fifth Limb: Comb with stout exopod bristle, 2 slender bristles ventral to base of exopod bristle, 5 short bristles close to ventral margin distal to base of exopod bristle, and 1 proximal and 1 or 2 distal bristles close to ventral margin. Epipod with 54 hirsute bristles. Oblique section with some bristles broken off, with 9 bristles along edge of remaining part and 1 bristle set back from edge; ventral margin of comb with 27 bristles. Approximate total bristles using an estimate of 18 bristles on oblique section based on number on adult female 109.

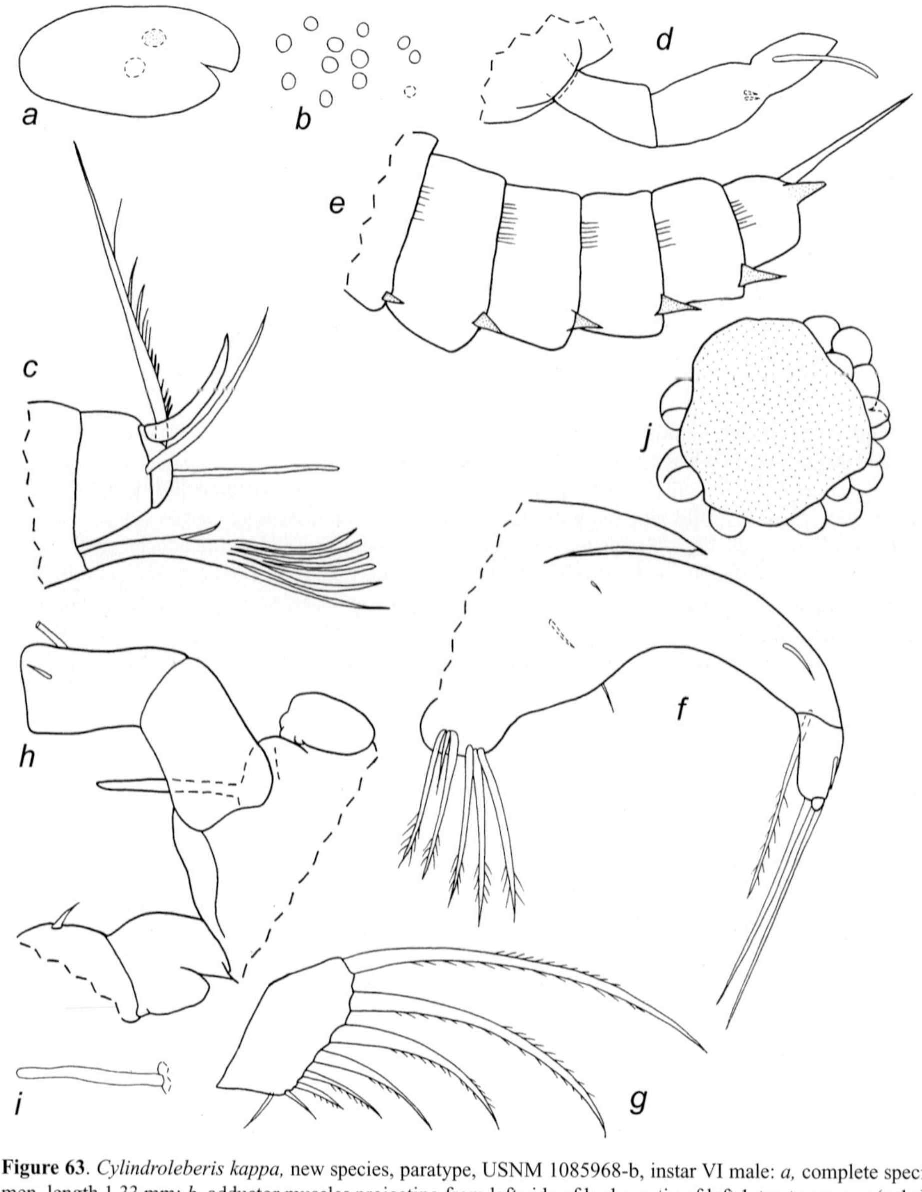

Figure 63. *Cylindroleberis kappa,* new species, paratype, USNM 1085968-b, instar VI male: *a,* complete specimen, length 1.33 mm; *b,* adductor muscles projecting from left side of body; *c,* tip of left 1st antenna, mv (nabs); *d,* endopod right 2nd antenna, mv; *e,* part exopod left 2nd antenna, lv (nabs); *f,* left maxilla, mv (nabs); *g,* right lamella of furca, lv; *h,* anterior of body from left side showing parts left 1st antenna and mandible, medial eye and Bellonci Organ; *i,* Bellonci Organ; *j,* lateral eye.

Sixth Limb: Shape similar to that of adult female. Anterior margin with 2 endites, each with 1 bristle. Anterior end of ventral margin with 5 bristles similar to those of adult female. Ventral and posteroventral margin with 13 bristles; bristles longer on posteroventral curvature. Anterior, ventral, and posterior margins hirsute. Total bristles 20.

Seventh Limb: Similar to that of adult female except bristles with slight taper. Total bristles 12.

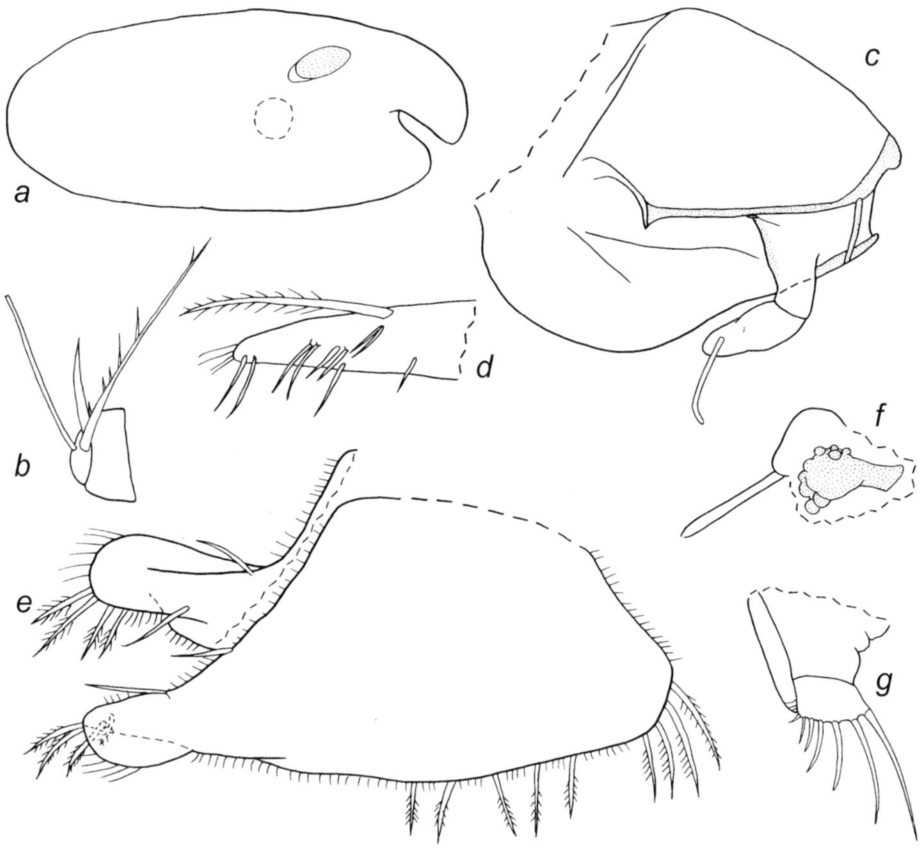

Figure 64. *Cylindroleberis kappa,* new species, paratype, USNM 1085968-d, instar V male: *a,* complete specimen, length 1.06 mm; *b,* tip left 1st antenna, lv (nabs); *c,* protopod and endopod left 2nd antenna, mv; *d,* comb left 5th limb, lv (nabs); *e,* left 6th limb (lv) and anterior right 6th limb, mv; *f,* medial eye, Bellonci organ, and left lateral eye; *g,* right lamella of furca, lv.

Furca (Figure 63*g*): Each lamella with 6 claws followed by 2 bristlelike claws. Total claws and bristles 8.

Bellonci Organ (Figure 63*h,i*): Similar to that of adult female.

Eyes: Medial eye similar to that of adult female (Figure 63*h*). Lateral eye with many ommatidia obscured by black pigment; eye slightly larger than medial eye (Figure 63*h, j*). Lateral eye larger than that of adult female.

Upper Lip: Not examined.

Anterior of Body (Figure 63*h*): Bare.

Posterior of Body: Similar to that of adult female.

Y-Sclerite: Similar to that of adult female.

Genitalia: None observed.

Gills: With 7 well developed gills overhanging posterior end of body, with convex tips.

Gut Content: Amber colored minute particulate matter.

Description of instar V male (Figure 64). Shape similar to that of adult female (Figure 64*a*).

Infold: Rostral infold with 23 bristles. Anteroventral infold with 15 bristles; 1 small bristle at inner end of incisure. Ventral infold with 11 bristles. Posterior infold with 5 processes near outer edge similar to those of adult female. Flaplike bristles obscured.

Selvage: Similar to that of A-1 male.

Central Adductor Muscle Attachments: Similar to those of adult female.

Carapace Size (length, height in mm): USNM 1085969, 1.06, 0.59.

First Antenna: Similar to that of adult female, except f-bristle with 3 proximal spines (Figure 64*b*).

Second Antenna (Figure 64*c*): Protopodite bare; sclerites in vicinity of endopod similar to that of adult female. Endopod with 3 articles: Article 1 bare; article 2 slightly longer than article 1, with 1 minute spinelike distal bristle; article 3 slightly shorter than article 2, with subterminal filament and rounded tip. Exopod with 9 articles; article 1 bare; bristle of article 2 just reaching article 8; bristles of articles 3 to 8 long with natatory hairs; 9th article with 4 bristles, 2 short bare (dorsal), 1 medium length bare, 1 long with natatory hairs. Basal spines on articles 4 to 8 and lateral spine of 9th article similar to those of A-1 male. Total bristles 14.

Mandible: Basis and exopod similar to those of adult female. Endopod: 1st article similar to that of adult female. 2nd article: ventral margin with 3 terminal bristles; dorsal margin with stout a-, b-, c-, and d-bristles, and 1 small bristle proximal to a-bristle; lateral side with long bristle between c- and d-bristles; medial side with 3 spinous bristles forming oblique row adjacent to c-bristle and 1 short bristle just distal to d-bristle. 3rd article similar to that of adult female. Total bristles 40.

Maxilla: Comb with 55 bristles plus proximal bent bristle. Limb otherwise similar to that of adult female. Approximate total bristles 71.

Fifth Limb (Figure 64*d*): Comb with stout spinous exopod bristle; 2 slender bristles ventral to base of exopod bristle; 8 bristles closer to ventral margin. Epipod with 49 bristles. Ventral margin of comb with 26 bristles; oblique section with 13 bristles. Total bristles 99.

Sixth Limb (Figure 64*e*): Anterior margin with upper and lower endite bristles. Anterior corner of ventral margin with 4 spinous bristles; posterior half of ventral margin with 9 spinous bristles; bristles longer on posteroventral curve. Total bristles 15.

Seventh Limb: Except for bristles being tapered and bearing 1 to 3 bells, limb similar to that of adult female. Total bristles 12.

Furca (Figure 64*g*): Each lamella with 6 claws followed by 1 bristlelike claw. Total claws and bristles 7.

Bellonci Organ (Figure 64*f*): Elongate with rounded tip.

Eyes (Figure 64*f*): Lateral eye with many ommatidia (about 19) obscured by black pigment. Medial eye bare, about same size as lateral eye, without pigment.

Lips: Similar to those of adult female.

Anterior of Body: Similar to that of adult female.

Posterior of Body: Similar to that of adult female, spinous.

Y-Sclerite: Similar to that of adult female.

Genitalia: Absent.

Gills: Similar to those of adult female.

Gut Content: Gut amber colored. Material darker and with unidentified black particles in anal area.

Description of instar IV male (Figures 65, 66*a–d*). Similar to that of adult female except for more oblique posterodorsal corner (Figure 64*a*).

Infold: Not examined in detail.

Selvage: Similar to that of instar V male.

Central Adductor Muscle Attachments (Figure 65*a,h*): About 8 ovoid attachments.

Carapace Size (length, height in mm): USNM 1085970, 0.83, 0.42.

First Antenna (Figure 65*b*): Article 1 bare. Article 2 with spinous dorsal bristle and small distal lateral bristle. Suture absent between articles 3 and 4; article 3 with small ventral bristle on inferred short ventral margin and 4 long bristles (proximal bristle with long spines) on long dorsal

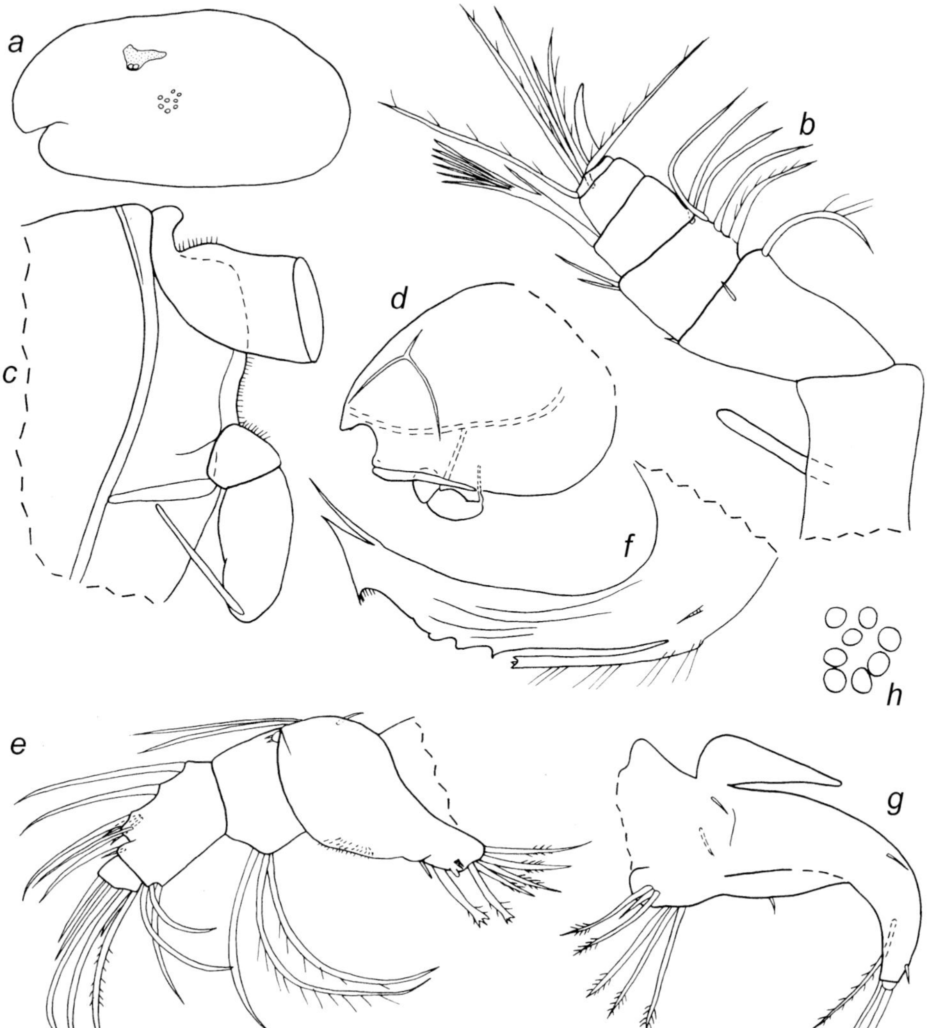

Figure 65. *Cylindroleberis kappa,* new species, paratype, USNM 1085970, instar IV male: *a,* complete specimen, length 0.83 mm; *b,* left 1st antenna (lv) and Bellonci Organ; *c,* left 2nd antenna, mv except twisted exopod (nabs); *d,* left 2nd antenna, lv; *e,* left mandible, lv; *f,* coxa endite left mandible, mv; *g,* left maxilla, mv (nabs); *h,* central adductor muscle protruding from left side of body.

margin; article 4 with 2 short ventral bristles and 1 long dorsal bristle. Sensory bristle of article 5 with 1 short proximal and 6 long terminal filaments. Articles 6 to 8 fused; article 6 with long medial bristle. Article 7: a-bristle clawlike; b- bristle with 3 marginal filaments; c-bristles with 5 marginal filaments. Article 8: d-bristle absent; e-bristle long, bare, with blunt tip; f-bristle bent dorsally with 3 proximal spines and 4 distal short filaments; g-bristle long with 5 short filaments. Total bristles 18.

Second Antenna: Protopod bare; sclerites in vicinity of endopod similar to those of adult female (Figure 65*c,d*). Endopod with 3 articles (2nd and 3rd articles fused); 2nd article slightly longer than

1st article, bare. 3rd article with long subterminal filament (Figure 65*c*). Exopod with 9 articles: bristle of article 2 reaching midlength of 7th article; bristles of articles 3 to 8 long with natatory hairs; bristles of articles 2 to 6 also with proximal slender ventral spines; 9th article with 3 bristles (2 short, 1 long (ventral) with natatory hairs). Articles 4 to 8 with short pointed basal spines; 9th article with pointed lateral spine about one-half length of article. Left limb of USNM 1085970 with spinelike foreign growth on proximal part of endopod, exopod, and adjacent protopod of left limb (Figure 65*c*), and endopod of right limb. Total bristles 11.

Mandible (Figure 65*e,f*): Coxa endite similar to that of adult female (Figure 65*f*). Basis endite with 2 triaenid bristles with 2 pairs of distal spines in addition to terminal pair, small process adjacent to 2 dwarf bristles, and 4 spinous end bristles. Basis with minute bristle close to endite, 2 long dorsal terminal bristles, and 1 short dorsal bristle near midlength. Exopod small with 2 small terminal bristles. Endopod: article 1 with 3 long ventral terminal bristles (2 with long marginal spines). Article 2: ventral margin with 3 long terminal bristles; dorsal margin with stout a-, b-, c-, and d-bristles, and 1 long lateral bristle between c- and d-bristles; medial side with 2 cleaning bristles adjacent to c-bristle, and 1 short bristle adjacent to d-bristle. Article 3 with dorsal claw, 3 long terminal bristles, 1 short medial bristle near ventral margin, and 1 long lateral bristle just ventral to dorsal claw. Ventral margin of basis of both limbs of USNM 1085970 with abundant spinelike structures (Figure 65*e*). Total bristles 37.

Maxilla (Figure 65*g*): Endite I with 2 long and 1 short bristle; endite 2 with 3 long bristles. Epipod extending to midlength of dorsal margin of basis. Basis with 1 proximal lateral bristle, 1 distal small ventral bristle, 1 long spinous terminal ventral bristle, 1 proximal bristle near dorsal margin, and 1 distal bristle near dorsal margin. Endopod: article 1 with short alpha-bristle and long beta-bristle; article 2 with long terminal bristle reaching past beta-bristle. Ventral margin of comb with 48 bristles plus proximal bent bristle. Total bristles 63.

Fifth Limb (Figure 66*a*): Epipod with 48 bristles; oblique section with 11 bristles along edge and 1 bristle set back from edge; ventral margin of comb with 17 bristles. Comb with stout spinous exopod bristle, 2 slender bristles just ventral to exopod bristle, and 5 bristles closer to ventral margin (4 near midlength, 1 proximal). Approximate total bristles 83.

Sixth Limb (Figure 66*c*): Anterior margin with bristle on upper and lower endites. Anterior corner of ventral margin with 3 spinous bristles; ventral margin with 2 or 3 spinous bristles; posteroventral curvature with 3 spinous bristles; lateral flap with hairs but no bristles; anterior, posterior and ventral margins hirsute. Total bristles 10 or 11.

Seventh Limb: Each limb with 6 proximal bristles, 3 on each side with 1 or 2 bells, and 4 distal bristles, 2 on each side with 1 or 2 bells; bristles strongly tapered. Terminus with opposing combs, each with 12 spinous teeth. Total bristles 10.

Furca (Figure 66*d*): Each lamella with 5 claws followed by 2 bristlelike claws. Total claws and bristles 7.

Bellonci Organ (Figures 65*b*, 66*b*): Elongate with rounded tip.

Eyes: Lateral eye with many ommatidia obscured by black pigment, height 0.054 mm (Figures 65*a*, 66*b*). Medial eye unpigmented, bare, about same size as lateral eye, with abundant short spinelike foreign structures along dorsal margin (Figure 66*b*).

Lips: Similar to those of adult female.

Anterior of Body: Similar to that of adult female.

Posterior of Body (Figure 66*d*): Similar to that of adult female, spinous.

Y-Sclerite (Figure 66*d*): Similar to that of adult female.

Genitalia: None.

Gills (Figure 66*d*): With 14 gills with rounded tips, 7 gills on each side.

Gut Content: Amber colored.

Foreign Organisms: Parts of 2nd antenna, ventral margin of basis of mandible, and dorsal margin of medial eye with short spinelike foreign structures (Figures 65*c,e*, 66*b*).

Figure 66. *Cylindroleberis kappa,* new species, paratype, USNM 1085970, instar IV, male: *a,* oblique row of bristles and comb of 5th limb; *b,* medial eye, left lateral eye, and Bellonci Organ; *c,* right 6th limb, lv; *d,* gills and furca projecting from carapace. Paratype, BPBM-S 12827, instar III, ?female: *e,* complete specimen, length 0.65 mm; *f,* right lamella of furca, lv; *g,* posterior of body showing Y-Sclerite and girdle. Paratype, USNM 1085965, instar III, ?female: *h,* complete specimen, length 0.68 mm; *i,* rostrum right valve, iv; *j,* anteroventral margin right valve, iv; *k,* posterior margin right valve, iv.

Description of instar III ?female (Figures 66*e–k,* 67). In general shape of carapace similar to that of adult female, but shell less sclerotized (Figure 66*e,h*).

Infold: Rostral infold with 13 bristle (Figure 66*i*). Anteroventral infold with 7 bristles (Figure 66*j*). Ventral infold with 5 bristles. Posterior infold with 5 small processes near outer edge and about 11 flaplike bristles and 3 or 4 minute bristles along inner edge (Figure 66*k*) .

Selvage: Similar to that of A-1 male.

Central Adductor Muscle Attachments: About 9 ovoid attachments.

Carapace Size (length, height in mm): USNM 1085965, 0.68, 0.35. BPBM-S 12827, 0.65, 0.32.

First Antenna (Figure 67*a*): 1st article bare. 2nd article with long spinous dorsal bristle and short distal lateral bristle. 3rd and 4th articles without separating suture. 3rd article with minute bristle on short ventral margin and 2 spinous bristles on long dorsal margin (proximal with long spines, distal with short spines). 4th article with 1 terminal ventral bristle and 1 terminal dorsal bristle with base medial; 4th and 5th articles separated by slightly concave suture on medial side and no suture on lateral side. Sensory bristle of 5th article with 1 short proximal and 6 long terminal filaments. 6th article with long medial bristle. 7th article: a-bristle clawlike, dorsal; b-bristle with 3 marginal filaments; c-bristle with about 5 marginal filaments. 8th article: d-bristle absent; e-bristle about one-half length of c-bristle, bare; f-bristle bent dorsally, with 4 marginal filaments; g-bristle with about 4 marginal filaments. Total bristles 15.

Second Antenna (Figure 67*b*): Protopod bare; sclerites near endopod similar to those of adult female. Endopod with single article with terminal filament. Exopod: 1st article bare; 2nd article with bristle reaching 8th article; articles 3 to 8 with long natatory setae; 9th article with 3 bristles (dorsal bristle short). Articles 2 to 8 with small triangular basal spines; 9th article with small triangular lateral spine. Total bristles 11.

Mandible (Figure 67*c–e*): Coxa endite similar to that of adult female, with small bristle near base of ventral branch (Figure 67*e*). Basis endite with 1 or 2 triaenid bristles with 2 pairs of marginal spines in addition to terminal pair, small process adjacent to 2 dwarf bristles, 3 spinous end bristles, and 1 minute bristle on endite near base (Figure 67*c,d*). Basis with 2 long terminal dorsal bristles, but no dorsal midbristle. Exopod small with 2 small terminal bristles. Endopod: 1st article with 3 long ventral bristles (2 with long spines). 2nd article: ventral margin with 3 terminal bristles; dorsal margin with a-, b-, c-, and d-bristles similar to those of adult female, 1 long lateral bristle between c- and d-bristles, 1 short spinous medial bristle adjacent to c-bristle, 1 short spinous medial bristle adjacent to d-bristle. 3rd article with stout dorsal clawlike bristle, 3 long end bristles, 1 short medial bristle near ventral margin, and 1 longer lateral bristle just ventral to clawlike bristle. Total bristles 31 or 32.

Maxilla (Figure 67*f*): Endite I with 2 long and 1 short bristle; endite II with 3 long bristles. Basis with small proximal dorsal bristle, proximal lateral bristle, small ventral bristle proximal to midlength, 1 long terminal spinous ventral bristle, and 1 distal bristle near dorsal margin. Endopod: 1st article with long beta-bristle (alpha-bristle not observed). 2nd article with terminal bristles slightly longer than beta-bristle. Ventral margin of comb with 43 bristles plus proximal bent bristle. Total bristles 55.

Fifth Limb: Comb with stout spinous exopod bristle reaching past tip of comb; 2 slender bristles ventral to base of comb; at least 2 additional bristles near ventral margin (obscured). Epipod with 43 bristles; oblique section with 10 bristles; ventral margin of comb with 16 bristles. Approximate total bristles 74.

Sixth Limb (Figure 67*g,h*): Anterior margin with upper and lower endite bristles; ventral margin with 2 spinous bristles at anterior tip, 1 ventral bristle posterior to midlength, and 2 longer bristles on posterodorsal corner. Anterior, ventral, and posterior margins hirsute. Total bristles 7.

Seventh Limb (Figure 67*i*): Elongate without bristles; tip rounded, bare.

Furca (Figure 66*f*): Each lamella with 4 claws followed by 2 bristles. Total claws and bristles 6.

Belloncí Organ (Figure 67*j*): Elongate with rounded tip.

Eyes: Lateral eye with ommatidia obscured by black pigment (Figures 66*e,h,* 67*k*). Medial eye unpigmented, about same size as lateral eye (Figure 67*j*).

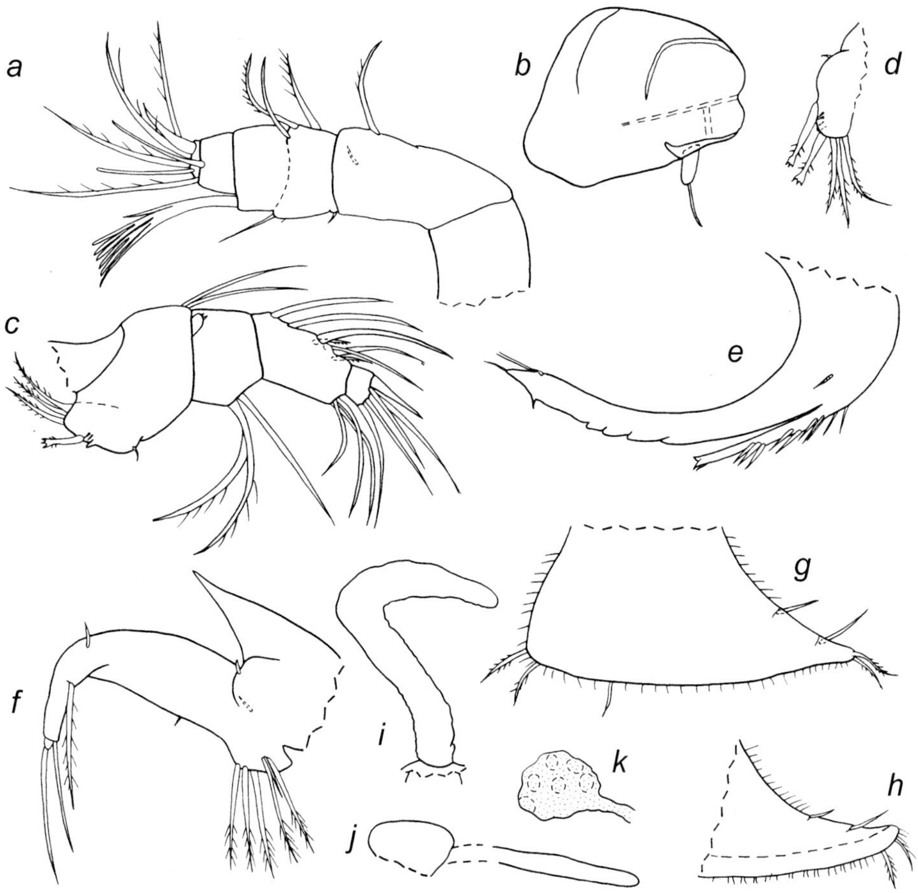

Figure 67. *Cylindroleberis kappa,* new species, paratype, USNM 1085967a, instar III, ?female: *a,* right 1st antenna, mv; *b,* protopod and endopod right 2nd antenna, lv; *c,* right mandible, lv; *d,* basis endite left mandible, lv; *e,* coxa endite left mandible, mv; *f,* right maxilla, mv; *g,* right 6th limb, lv; *h,* anterior part left 6th limb, mv; *i,* 7th limb; *j,* medial eye and Bellonci Organ; *k,* left lateral eye.

Upper Lip: Similar to that of adult female.
Anterior of Body: Similar to that of adult female.
Posterior of Body: Similar to that of adult female.
Y-Sclerite (Figure 66g): Similar to that of adult female.
Genitalia: Not present.
Gills: Similar to those of adult female.
Gut Content: Amber-colored.

Remarks. Specimen tentatively identified as female because of unjointed endopod of 2nd antenna and lack of proximal spines on f-bristle of 1st antenna. The "Generalized Key to Early Instars of Cylindroleberididae" in Kornicker (1981: 38) identifies the specimen as an Instar III.

Comparisons. The maxilla of the new species *C. kappa* differs from that of *C. variabilis* Kornicker in having 2 instead of 3 long bristles on endite I. The maxilla of *C. nodulifera* (Poulsen) has 2 long

Table 6. Comparison of selected morphological characters on stages III to adult of *Cylindroleberis kappa* (ant. = anterior, art. = article, brist. = bristles, dors. = dorsal, f = female, m = male, na = not applicable).

Character	Instar III ?f	Instar IV m	Instar V m	Instar VI m	Adult f(m)
Carapace length (avg. mm)	0.67	0.83	1.06	1.32	1.27(1.50)
First antenna (bristles)					
3rd article (dorsal)	2	4	6	6	6(6)
4th article (ventral)	1	2	2	2	2(2)
Second antenna (bristles)					
Endopod male 2nd article	na	0	1	2	na(3)
Exopod 9th article	3	3	4	4	4(4)
Mandible (bristles)					
Basis middle dorsal margin	0	1	1	1	1(1)
Sixth limb (bristles)					
anterior tip	2	3	4	5	5(5)
ventral margin	3	5–6	9	13	11–12(15)
Seventh limb (bristles)	0	10	12	12	12(12)
Furca (claws + bristles)	6	7	7	8	9(9)

Table 7. Approximate total number of bristles on appendages of stages III to adult of *Cylindroleberis kappa* (f = female, m = male).

Appendage	Instar III f	Instar IV m	Instar V m	Instar VI m	Adult f(m)
First antenna	15	18	20	20	20(25)
Second antenna	11	11	14	15	13(16)
Mandible	32	37	40	41	40–43(42–43)
Maxilla	55	63	71	74	77(74)
Fifth limb	74	83	99	ca. 109	104(109)
Sixth limb	7	11	15	20	18(22)
Seventh limb	0	10	12	12	12(12)
Furca	6	7	7	8	9(9)
Total	200	240	278	299	296(310)

bristles on endite I of the maxilla, but the mandible is without a bristle near the middle of the dorsal margin of the basis of the mandible. *C. rangiroaensis* Hartmann differs from *C. kappa* in having a smaller and less elongate carapace, and in not having a small proximal filament on the sensory bristle of the 5th article of the female 1st antenna.

Ontogeny and sexual dimorphism. *C. kappa* has 6 instars plus the adult, and the latter does not molt. Only the last 5 stages are in the present collection. In general, the number of bristles on appendages tend to increase in later stages (Tables 6, 7). The average lengths of carapaces are given in Tables 6 and 8, and growth factors in Table 8. The number of bristles on the infold of the carapace was counted in only a few stages, but the number of bristles tends to increase in later stages. The small processes along the posterior part of the posterior infold of the carapace are already present on Instar III. The posterior part of the dorsal margin is oblique on Instars III and IV. The f-bristles on the 1st antennae of male instars IV, V, and VI bear several short spines proximal to the usual short filaments. Bristles first appear on the 7th limb in Instar IV, which is typical for the Myodocopina (Kornicker & Harrison-Nelson, 1999: 35). The bristles of the 7th limb of the adult are either cylin-

Table 8. Average shell length and calculated growth factors for *Cylindroleberis kappa* (f = female, m = male).

Growth stage	Average length (mm)	Growth factor	Number of specimens
Instar III (f)	0.67		2
		1.24	
Instar IV (m)	0.83		1
		1.28	
Instar V (m)	1.06		1
		1.25	
Instar VI (m)	1.32		1
		1.14	
Adult (m)	1.50		1
Average		1.23	

drical or slightly tapered, whereas those of instars IV–VI are tapered, with taper decreasing on later instars. The bristles on instar VI are only slightly tapered and not greatly different from those on the adult.

Observed sexual dimorphisms of appendages of *C. kappa* are similar to those observed previously on members of the family (Kornicker, 1981: 30; Poulsen, 1965: 478); dimorphisms of the 1st and 2nd antennae of *C. kappa* are shown herein in the illustrations of those appendages. The carapace of the adult male *C. kappa* is larger than that of the adult female (adult male length 1.50 mm (N = 1); female length 1.22–1.37 mm (N = 5)). The adult male carapace bears the usual vertical row of bristles near the posterior margin and the rostrum projects farther forward than on the adult female. The basis endite of the adult male mandible does not have the triaenid bristles present on both the adult female and juveniles of both sexes; instead, ordinary bristles are present.

Fecundity. Some adult females have small unextruded eggs in addition to large eggs in the marsupium showing that they have more than one clutch of eggs. Many adult females in the collection bear eggs in the marsupium. The number of eggs in the marsupia of 6 specimens ranged from 4 to 21, average 10. The observed variability is not unusual in myodocopids (Kornicker & Harrison-Nelson, 1999: 40).

LITERATURE CITED

Brady, **G.S**. 1867. A synopsis of Recent British Ostracoda. *Intellectual Observer* **12**(2): 110–130.

———. 1898. On new or imperfectly known species of Ostracoda, chiefly from New Zealand. *Transactions of the Royal Society of Edinburgh* **14**(8): 429–452.

Cohen, **A.C. & Kornicker**, **L.S. 1975**. Taxonomic indexes to Ostracoda (Suborder Myodocopina) in Skogsberg (1920) and Poulsen (1962, 1965). *Smithsonian Contributions to Zoology* **204**, 29 p.

DeFelice, **R.C. & Minton**, **D**. 2001. Marine species survey of Johnston Atoll, Central Pacific Ocean, June 2000. *Bishop Museum Technical Report* **19**

Coles, **S.L.**, **DeFelice**, **R.C. & Eldredge**, **L.G**. 2002a. Nonindigenous marine species introductions in Kane'ohe Bay, Oahu, Hawai'i. *Bishop Museum Technical Report* **24**, 353 p.

———., **DeFelice**, **R.C. & Eldredge**, **L.G**. 2002b. Nonindigenous marine species introductions at Waikiki and Hawai'i Kai, O'ahu, Hawai'i. *Bishop Museum Technical Report* **25**, 244 p.

DeFelice, **R.**, **Minton**, **D. & Godwin**, **S**. 2002. Records of shallow-water marine invertebrates from French Frigate Shoals, Northwestern Hawaiian Islands with a note on nonindigenous species. *Bishop Museum Technical Report* **23**, 78 p.

Hartmann, G. 1984. Zur Kenntnis der Ostracodender polynesischen Inseln Huahine (Gesellschaft-inseln) und Rangiroa (Tuamotu-Inseln). *Mitteilungen aus dem Hamburgischen Zoologischen Museum und Institut* **81**: 117–169.

———. 1991. Ostracoden von Hawaii, inbesondere aus dem marinen Interstitial. *Helgolaender Meeres Untersuchungen* **45**: 165–198.

Kornicker, **L.S**. 1970. Ostracoda (Myodocopina) from the Peru-Chile Trench and the Antarctic Ocean. *Smithsonian Contributions to Zoology* **32**, 42 p.

———. 1975. Antarctic Ostracoda (Myodocopina) parts 1 and 2. *Smithsonian Contributions to Zoology* **163**, 720 p.

———. 1976. Benthic marine Cypridinacea from Hawaii (Ostracoda). *Smithsonian Contributions to Zoology* **231**, 24 p.

———. 1978. *Harbansus,* a new genus of marine Ostracoda, and a revision of the Philomedidae (Myodocopina). *Smithsonian Contributions to Zoology* **260**, 75 p.

———. 1981. Revision, distribution, ecology, and ontogeny of the ostracode subfamily Cyclasteropinae (Myodocopina: Cylindroleberididae). *Smithsonian Contributions to Zoology* **319**, 548 p.

———. 1986. Cylindroleberididae of the Western North Atlantic and the Northern Gulf of Mexico, and zoogeography of the Myodocopina (Ostracoda). *Smithsonian Contributions to Zoology* **425**, 139 p.

———. 1987. Supplementary description of *Cypridina americana* (Müller, 1890), a luminescent myodocopid ostracode from the East Pacific. *Proceedings of the Biological Society of Washington,* **100**(1): 173–181.

———. 1988. Chapter 21. Ostracoda (Myodocopina) of Enewetak Atoll, p. 217–219. *In*: Devaney, D.M., Reese, E.S., Burch, B.L. & Helfrich, P., eds., *The natural history of Enewetak Atoll. Vol. II. Biogeography and systematics,* U.S. Department of Energy, Office of Scientific and Technical Information, Oak Ridge, Tennessee.

———. 1991. Myodocopid Ostracoda of Enewetak and Bikini Atolls. *Smithsonian Contributions to Zoology* **505**, 140 p.

———. 1992. Myodocopid Ostracoda of the Benthedi Expedition, 1977, to the NE Mozambique Channel, Indian Ocean. *Smithsonian Contributions to Zoology* **531**, 243 p.

———. 1995. Ostracoda (Myodocopina) of the SE Australian Continental Slope, Part 2. *Smithsonian Contributions to Zoology* **562**, 97 p.

Kornicker, **L.S.**, **& Caraion**, **F.E**. 1974. West African Myodocopid Ostracoda (Cylindroleberididae). *Smithsonian Contributions to Zoology* **179**, 78 p.

———. **& Caraion**, **F**.**E**. 1977. West African myodocopid Ostracoda (Cypridinidae, Philomedidae), *Smithsonian Contributions to Zoology* **241**, 100 p.

———. **& Harrison-Nelson**, **E**. 1999. Eumeli Expeditions, Part 1: *Tetragonodon rex,* new species, and general reproductive biology of the Myodocopina. *Smithsonian Contributions to Zoology* **602**, 55 p.

———. **& Harrison-Nelson**, **E**. 2005. Ostracoda (Myodocopina) from Johnston Atoll, Pacific Ocean, and proposal of a new tribe Bruuniellini (Myodocopina: Cylindroleberididae). *Pacific Science* **59**(3): 323–362.

———. **& Harrison-Nelson**, **E**. 2006. Ostracoda (Myodocopina) of Tutuila, American Samoa. *Pacific Science* **60**: 243–259.

———. **& Thomassin**, **B**.**A**. 1998. Ostracoda (Myodocopina) of the Tuléar Reef Complex, SW Madagascar. *Smithsonian Contributions to Zoology* **595**, 134 p.

Milne-Edwards, **M**. 1840. Ordre des cyproides ou des ostracodes. *Histoire Naturelle des Crustacés* **3**: 393–411. Paris

Müller, **G**.**W**. 1890. Neue Cypridiniden. *Zoologische Jahrbücher* **5**: 211–252.

———. 1906. Ostracoda. *In*: Chun, C., ed., *Wissenschaftliche-Ergebnisse der Deutsche Tiefsee-Expedition auf dem Dampfer 'Valdivia' 1898–1899,* **8**(2): 1–154.

———. 1912. Ostracoda. *Das Tierreich* **31**, 434 p.

Poulsen, **E**.**M**. 1962. Ostracoda-Myodocopa, 1: Cypridiniformes-Cypridinidae. *Dana Report* **57**, 414 p.

———. 1965. Ostracoda-Myodocopa, 2: Cypridiniformes-Rutidermatidae, Sarsiellidae and Asteropidae. *Dana Report* **65**, 484 p.

Ross, **H**.**H**. 1974. *Biological systematics*. Addison-Wesley, Reading, Massachusetts.

Sharpe, **R**.**W**. 1908. A further report on the Ostracoda of the United States National Museum. *Proceedings of the United States National Museum* **35**(1651): 399–430.

Skogsberg, **T**. 1920. Studies on marine ostracods I: Cypridinids, halocyprids, and polycopids. *Zoologiska Bidrag från Upsala, Supplement* **1**: 1–784.

Sohn, **I**.**G**. **& Kornicker**, **L**.**S**. 1969. Significance of calcareous nodules in myodocopid ostracod carapaces. p. 99–108. *In*: Neale, J.W., ed., *The taxonomy, morphology and ecology of Recent Ostracoda.* Oliver & Boyd, Edinburgh.

Sylvester-Bradley, **P**.**C**. 1961. Myodocopida, p. (Q)387–406. *In*: Moore, R.C. ed., *Treatise on Invertebrate Paleontology* 3. Geological Society of America and University of Kansas Press.

Tseng, **W**.-**Y**. 1977. Pelagic Ostracoda of Taiwan. Part 1. Cypridiniformes. *Laboratory of Marine Resources, Taiwan Fisheries Research Institute, Report* **30**, 240 p.

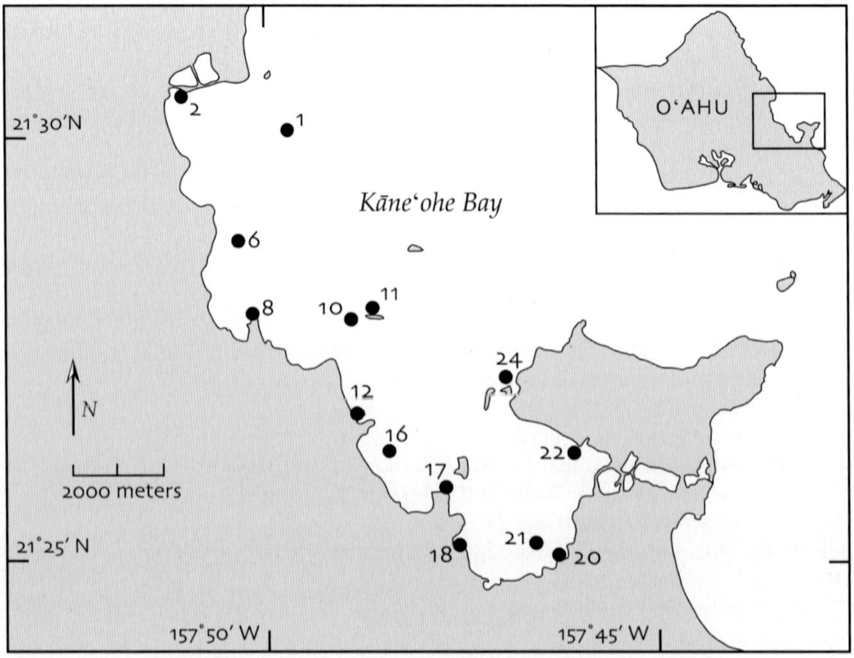

Map 1. Kāne'ohe Bay, showing location of stations where specimens were collected.

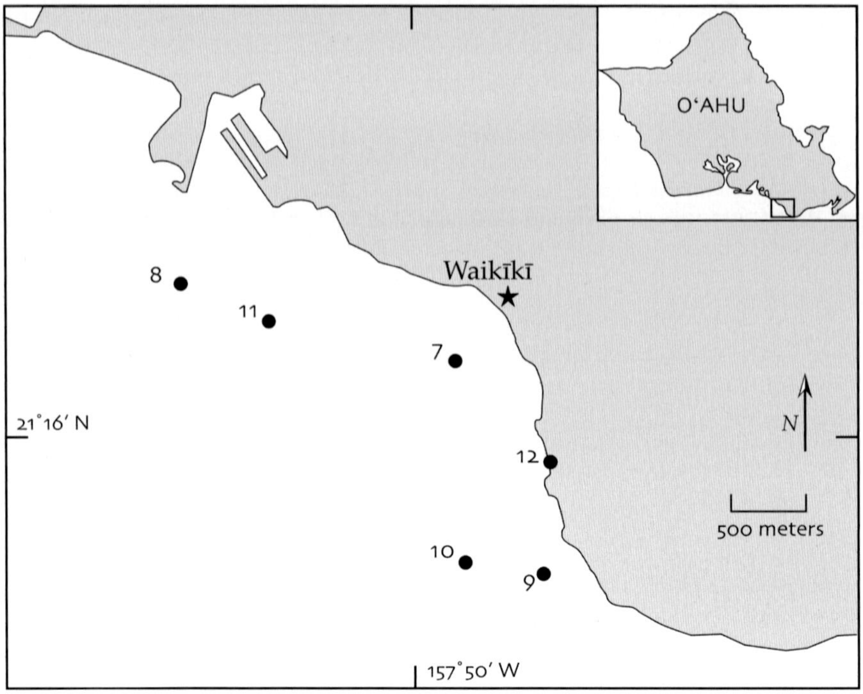

Map 2. Waikīkī, showing location of stations where specimens were collected.

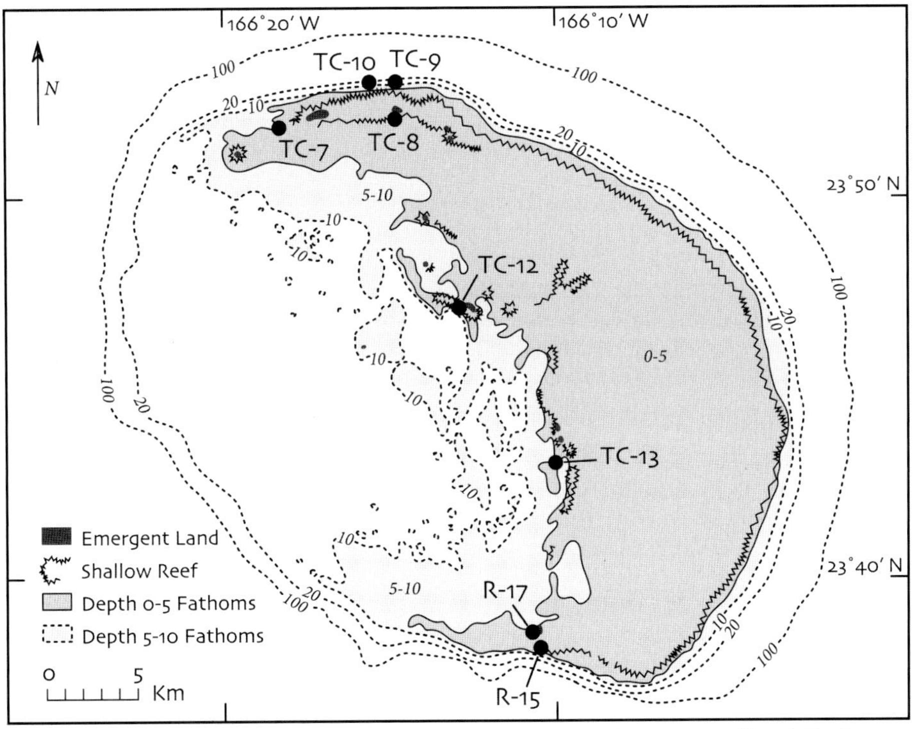

Map 3. French Frigate Shoals, showing location of stations where specimens were collected (R= *Rapture*, TC = *Townsend Cromwell*).

APPENDIX 1

Station Data and Number of Specimens of each Species Collected at O'ahu and French Frigate Shoals (see Maps 1–3).

Kāne'ohe Bay, O'ahu.

Sta 1, 17 Nov 1999, 21°30'22.1"N, 157°50'57.1"W, North Channel, depth 1–6 m. Wave turbulent area along south side of main channel near Buoy Marker 5, on reef area with abundant live coral extending from 4 to 6 m depth next to dredged channel. Outside of live coral area bottom is consolidated limestone with coarse sand and moderate *Halimeda* algae cover.

Parasterope sigma: 1 specimen.

Parasterope omega: 2 specimens.

Sta 2, 2 Nov 1999, 21°30'22.1"N, 157°44'50.6"W, Moli'i Fishpond, depth 0.5–1 m. Inshore area at north end of bay with abundant *Rhizophora* mangle mangroves growing along a muddy shore. Subtidal substratum is coarse sand with surface coating of fine, terrigenous silt. Abundant biotope on bottom is mats of nonindigenous macroalgae *Acanthophora spicifera* and *Gracilaria salicornia* mixed with *Hypnea musciformis*, *Kappaphycus* spp. and native *Caulerpa taxifolia* and *Dictyota acutiloba*. Burrows, probably of *Alpheus mackayi* and *Callianassa* sp. and tubes of the nonindigenous polychaete *Sabellastarte spectabilis* were common.

Cylindroleberis kappa: 8 specimens.

Sta 6, 19 Nov 1999, 21°28'39.7"N, 157°50'12.2"W, Pu‘u Kaua‘i reef, depth 0.5–6 m. Fringing reef extending from flat at 0.5 m to silty bottom at reef base in 6 m depth. Reef flat has coral cover of ca. 25%, mostly of *Porites compressa* and *Pocillopora damicornis*, with common calcareous *Hydrolithon* sp. and green *Dictyosphaeria cavernosa* and occasional *Gracilaria salicornia* algae and *Montipora capitata* coral. Sand channels are interspersed on reef top with medium to coarse sand. Reef slopes gently with high coverage of mostly *P. compressa* to start of fine silt sediment covered bottom at ca 6 m.

Parasterope omega: 2 specimens.

Sta 8, 12 Nov 1999, 21°27'48.7"N, 157°50'01.5"W, Kahalu‘u Landing pier, depth 0–2.5 m. Old pier pilings extending 200 m out from Wailau Point supporting fouling community. Fine silt bottom in 2.5 m depth shows abundant burrow openings and growth of seagrass *Halophila hawaiana*.

Paravargula trifax: 3 specimens.

Cypridina alpha: 1 specimen

Sta 10, 12 Nov 1999, 21°27'46.7"N, 157°48'45.9"W, Wass Reef, depth 1–9 m. Small patch reef near the Kāne‘ohe Bay sand bar that was the site of 2 comprehensive fish surveys utilizing total recovery by fish poisoning. Reef is steeply sloped from 1–9 m depth, with heavy *Dictyosphaeria cavernosa* coverage from the reef top to 7 m depth.

Cypridina alpha: 5 specimens.

Sta 11, 12 Nov 1999, 21°27'53.91"N, 157°48'31.6"W, Sand Bar, depth 0–0.5 m. Subtidal zone of ca. 0.5 m depth sampled about 200 m from lagoon side. Substratum is coarse sand and fine coral rubble with abundant seagrass *Halophila hawaiana* interspersed with dense patches of introduced algae *Acanthophora spicifera* and less abundant *Gracilaria salicornia*, *Dictyosphaeria cavernosa*, *Hypnea musciformis*, *Caulerpa taxifolia*, *Padina* sp., *Halimeda* sp. with clumps of broken coral *Montipora capitata*.

Cylindroleberis kappa: 4 specimens.

Sta 12, 19 Jan 2000, 21°26'36.8"N, 157°48'40.3"W, He‘eia Kea Pier, depth 0–7 m. Northwest side of main He‘eia pier and on a shallow reef flat separated from dock by a narrow channel. Pier pilings only sparsely covered with fouling organisms, mostly filamentous algae. Channel has abundant algae on bottom, and reef flat is covered with coarse sand and heavy growths of algae, especially introduced *Gracilaria salicornia*, *Hypnea musciformis*, *Kappaphycus* spp. and *Acanthophora spicifera*.

Cylindroleberis kappa: 1 specimen.

Sta 16, 10 Nov 1999, 21°26'13.2"N, 157°48'15.2"W, He‘eia Fish Pond Reef, depth 0.5–2.0 m. Reef outside of one of the largest former fishponds in the bay, which is now filled with sediment and lined with introduce *Rhizophora* mangle mangrove trees along much of its perimeter. Large sandy area on reef top in 0.5 m has intermittent coral rubble and abundant introduced *Gracilaria salicornia* algae. Reef slopes down to 2 m with moderate cover of *Porites compressa* and *Montipora capitata* coral and abundant *Dictyosphaeria cavernosa* and common *Kappaphycus* spp.

Paravargula trifax: 1 specimen.

Sta 17, 8 Nov 1999, 21°25'46.9"N, 157°47'31.1"W, Hawaii Institute of Marine Biology pier, depth 0–4 m. Floating docks and cement pilings for the HIMB ferry landing, built approximately 200 m out onto the edge of the fringing reef. Reef flat at 0.5–1.0 m depth is fine to coarse sand and coral rubble on limestone substratum, with abundant *Gracilaria salicornia*, *Acanthophora spicifera* and *Padina* sp. algae. Pier pilings and floating docks support abundant fouling invertebrates dominated by the introduced sponge *Mycale armata*, and corals *Porites compressa*, *Montipora capitata* and *Pocillopora damicornis* are common to abundant down slope from reef edge.

Cylindroleberis kappa: 46 specimens.
Cypridina alpha: 1 specimen.

Sta 18, 8 Nov 1999, 21°25'06.0"N, 157°47'23"W, Makani Kai Marina, depth 0–3 m. Floating docks and concrete wall in a harbor area nearly isolated from the bay by a rock jetty and much affected by fresh water runoff. Introduced oysters *Crassostrea* sp., fanworm *Sabellastarte spectabilis* and bryozoan *Amathia distans* are abundant along the concrete wall along with numerous other fouling invertebrates. Few macro-organisms were seen on the floating docks. Only a single coral species *Montipora capitata* occurred at 2 m depth, well below freshwater lens.

Paravargula trifax: 1 specimen.

Sta 20, 5 Nov 1999, 21°25'2.6"N, 157°46'7.2"W, Kāne'ohe Yacht Club dock, depth 0–6.5 m. Pier pilings and floating docks of the Kāne'ohe Yacht Club, the largest marina in the bay, which provides berthing to both motor and sailing craft. Channel entrances to the yacht club and its basins have been dredged from the reef to ca. 7 m depth. Docks and piling support an abundant fouling community, dominated by introduced species, especially the bryozoans *Schizoporella* sp. and *Amathia distans*, the polychaete *Sabellastarte spectabilis*, and a few *Porites compressa* and *Pocillopora damicornis* corals occurred.

Paravargula trifax: 3 specimens.

Sta 21, 5 Nov 1999, 21°25'09.5"N, 157°46'22.6"W, Kāne'ohe Yacht Club patch reef, depth 1–5 m. Located between entrance channels to the Yacht Club, the reef extends from 0.5 m down to 10 m depth. Reef top is mixed *Montipora capitata* and *Porites compressa* corals in coarse sand with abundant macroalgae *Gracilaria salicornia*, *Dictyosphaeria cavernosa* and *Kappaphycus* spp. Introduced invertebrates, e.g. the polychaete *Sabellastarte spectabilis*, the ascidian *Phallusia nigra* and the hydroid *Pennaria disticha* were abundant, along with the synaptid holothurian *Opheodesoma spectabilis*.

Paravargula trifax: 1 specimen.
Cypridina alpha: 1 specimen.

Sta 22, 25 Jan 2000, 21°26'15.3"N, 157°45'55.1"W, Marine Corps Base Fuel Dock, depth 0.5–8.0 m. Concrete and iron pilings of the main docking area for the Kāne'ohe Marine Base on the southwest side of Mōkapu Peninsula and east of the base runways. Surfaces are densely fouled with heavy coverage of sponges and other invertebrates to the bottom in 9 m depth. Sponges are sediment coated and bottom is silty fine sediment with numerous borrow openings

Paravargula trifax: 1 specimen.

Sta 24, 25 Jan 2000, 21°27'03.03"N, 157°46'42"W, SAG Harbor, Marine Corps Base, depth 0.5 8.0 m. Docking area from small to medium size boats along northwest shore of Mōkapu Peninsula with channel and basin dredged to 8 m depth. Area is within 100 m of a reef with high coral coverage, including the species *Porites compressa*, *Montipora capitata*, *Pocillopora damicornis* and *Fungia scutaria*, and an abundance of a variety of reef fishes. The pier pilings support numerous introduced fouling organisms including a heavy coverage of the nonindigenous octocoral *Carijoa riisei*.

Cypridina alpha: 1 specimen.

Waikīkī, Oahu.

Sta 7, 18 Jan 2001, 21°16'18.3"N, 157°49'42.5"W, Canoes, depth 3.5 m. Outside of "Canoes" surf site in 3.5 m depth. Barren, sand scoured reef with heavy *Dictyopteris australis* coverage and very little coral, surrounded by sand with much sand accumulated on reef. High water turbidity at time of survey from surf-suspended sand.

Parasterope iota: 2 specimens.
Cypridina alpha: 9 specimens.

Sta 8, 18 Jan 2001, 21°16'34.0"N, 157°50'46.7"W, Ala Wai Buoy, depth 16.5 m. Ala Wai entrance buoy #2 anchored in 16.5 m depth with very little fouling on buoy and none on chain. Bottom substratum at buoy is a hard and smooth fossil reef with light fine sand; cobble size rubble with small outcrops increases with distance away from the buoy anchor site.

Parasterope zeta: 3 specimens.
Cypridina alpha: 20 specimens.
Cypridina iota: 2 specimens.

Sta 9, 22 Jan 2001, 21°15'32.9"N, 157°49'21.6"W, Elks Club, depth 1.5 m. Shallow reef area off Elks Club in 1.5 m depth with flat coarse sand bottom and cobbles interspersed among dead reef.

Cypridina alpha: 3 specimens.

Sta 10, 22 Jan 2001, 21°15'34.8"N, 157°49'39.7"W, Kapua Channel, 5 m. Flat, low relief line of reefs among sand channels in 5 m depth; reefs highly scoured with light sand cover; abundant *Dichtyopteris australis* and sparse *Pocillopora meandrina* with ca. 1% cover.

Parasterope gamma: 1 specimen.
Parasterope iota: 1 specimen.
Parasterope zeta: 1 specimen.
Synasterope delta: 1 specimen.
Cypridina alpha: 51 specimens.

Sta 11, 22 Jan 2001, 21°16'26.5"N, 157°50'26.1"W, Kaiser's Channel, depth 10–12 m. East side of channel on reef with moderate (ca. 10%) coral cover with *Porites lobata* heads to 1.5 m diameter in 10 m depth on reef, to 12 m in sand channel with the highest coral cover and most pristine conditions of any site on the Waikīkī survey.

Parasterope zeta: 6 specimens.
Parasterope theta: 1 specimen.
Cypridina alpha: 18 specimens.

Sta 12, 23 Jan 2001, 21°15'57.2"N, 157°49'20.3"W, Aquarium Reef, depth 0–5 m. Reef area immediately offshore of Waikīkī Aquarium. Substratum is a shallow consolidated limestone reef 0–0.25 m deep from the beach to 25 m from shore that is separated from an offshore reef by a sand channel ca. 5 m deep. The Aquarium discharge pipe extends across the inshore reef to the channel and discharges warm effluent at ca. 3 m depth. On the nearshore reef coral is very scarce and macroalgae is very abundant and dominated by nonindigenous *Gracilaria salicornia* and *Acanthophora spicifera*, which accumulate in large deposits along the shoreline seawall. On the offshore reef intermittent corals, mostly *Porites lobata* and *Pocillopora meandrina*, occur, but the substratum is still virtually covered by a variety of macroalgal species, including *G. salicornia* and *A. spicifera* to 300 m offshore, after which the sandy bottom becomes dominated by *Dictyopteris australis*.

Parasterope omega: 1 specimen.

French Frigate Shoals
Specimens collected by R. DeFelice (FFS-R) and D. Minton (FFS-TC).
(Research Vessels: R = Rapture; TC = Townsend Cromwell)

FFS-TC-7, 11 Sep 2000, 23°52.0'N, 166°18.28'W, forereef, 2 km W of Tern Island, depth 5–10 m. Spur and groove habitat with very little current and surge. Spurs abut 5 m high and 5 m wide with encrusting *Porites lobata* dominating upper exposed surfaces. Undercut spur base shows evidence of occasional scour from intermittent surge. Groove floors at 10 m depth about 5 m wide. Groove channels with abundant carbonate sediments and coral rubble.
Cypridina alpha: 3 specimens.

FFS-TC-8, 12 Sep 2000, 23°52.494'N, 166°15.032'W, N fore reef, N of Trig Island, depth 10–17 m. Spur and groove habitat with weak current and low surge at time of dive. Spurs massive and with moderate coral cover, mainly encrusting *Porites lobata*. Spur slopes steep and groove floors with carbonate sediments and rubble. Boring urchins, especially *Echinostrephus acciculatus*, numerous on upper exposed surface of spurs.
Harbansus hox: 1 specimen.
Cypridina alpha: 26 specimens.

FFS-TC-9, 12 Sep 2000, 23°52.9'N, 166°15'W, N forereef, N of Tern Island, depth 10–15 m. Spur and groove habitat with some surge. Spurs of moderate height (4 m) and width (4 m), with low coral cover (encrusting *Porites lobata*) and considerable bare consolidated reef surfaces; evidence of frequent scour.
Cypridina alpha: 1 specimen.

FFS-TC-10, 12 Sep 2000, 23°52.9'N, 166°15.8'W, N forereef, W of Tern Island, near Shark Island, depth 3–10 m. Similar spur and groove habitat as FFS-TC-7, but consolidated reef with relatively lower relief and more discontinuous, separated by shallow sand/rubble grooves and patches. Surge low at time of dive, but unconsolidated sediment harbored few macroinvertebrates, which suggests this area frequently disturbed by ocean swell. Encrusting *Porites lobata* most abundant coral.
Cypridina alpha: 3 specimens.

FFS-TC-12, 13 Sep 2000, 23°47.326'N, 166°13.198'W, Central lagoon patch reef, 1 km E of East Island, depth 5–15 m. Patch reef with base at 15 m, surrounded by coarse sand flats with some rubble fields and isolated deposits. *Porites lobata* most abundant coral on some of the upward facing surfaces. Reef sides heavily eroded by unusually dense population of *Arca ventricosa* with some *Chama iostoma*, forming a virtual "bivalve reef".
Parasterope gamma: 1 specimen.
Cypridina alpha: 1 specimen

FFS-TC-13, 14 Sep 2000, 23°43.957'N, 166°10.375'W, Central lagoon reef, 0.5 km W of Gin Island, depth 3–10 m. Isolated patch reef with base at about 10 m depth surrounded by flats of shifting coarse sediment with some rubble deposits. Patch reef heavily eroded, with abundant turf algae and blue-greens. Top of reef with encrusting *Porites compressa*.
Parasterope zeta: 1 specimen.
Cypridina alpha: 2 specimens.
Harbansus hox: 8 specimens.

FFS-R-15, 23 Sept 2000, 23°38.530'N, 166°10.427'W, 0.5 km S of Disappearing Island, depth 2–7 m. Shallow forereef flat with substrate predominantly unconsolidated and poorly sorted reef rubble deposits mixed with carbonate sediments. Small patches and mounds of uneven unconsolidated substrate harbor the sparse coral at this site, predominantly *Pocillopora meandrina*. A strong current from east and surge from south present at time of dive and suspended sediment was present in water column. Many large corals broken or overturned suggesting recent or current disturbance from large waves striking reef from the south.

Cypridina alpha: 1 specimen.

FFS-R-17, 23 Sep 2000, 23°38.738'N, 166°10.468'W, 0.5 km W of Disappearing Island, depth 2–7 m. Shallow reef passage and backreef pool with large heads of *Porites lobata* surrounded by deep rippled sand bottom. Site appeared to be protected from impact of chronic ocean swell and recent storms. Coral cover high on reef mounds and with 2 table corals: *Acropora valida* and *A. cytherea*.

Cypridina alpha: 1 specimen.

APPENDIX 2

Species of *Harbansus* with or without lateral eyes and natatory bristles on the exopod of the 2nd antenna (F = adult female, M = adult male; III or IVM or F = Instars III or IV male or female; - = without, + = with) [in part from Kornicker (1992: Table 12)].

Species	With (+) or without (-) natatory bristles	Depth (meters)	Lateral eyes
H. barnardi Kornicker, 1978	- F, IVM	2	+
H. bowenae Kornicker, 1978	- F, IVF, IVM; + M	42–196	-
H. bradmeyersi Kornicker, 1978	- F; + M	11–60	+
H. dayi Kornicker, 1978	-F, IIIF, IIIM, IVM; +M	42–198	-F, IIIF; +IIIM, IVM, M*
H. felix Kornicker, 1995	- F	204–400	-
H. ferox Kornicker, 1992	- F	450	-
H. flax Kornicker & Thomassin, 1998	+ IVF	ca 2	+
H. hapax Kornicker, 1995	- F	429–996	-
H. hox Kornicker *et al.*, n. sp.	+ F, IVF, IVM; - IIIF	shallow	+
H. magnus Kornicker, 1984	- IVF	460	-
H. mayeri Kornicker, 1978	- F	8	+
H. paucichelatus (Kornicker, 1958)	+ F, M, I, IIIF, IIIM, IVF, IVM	1–35	+
H. rhabdion (Kornicker, 1970)	- F	991–1015	-
H. schornikovi (Kornicker & Caraion, 1977)	- IVM, IIIF	260	+**
H. slatteryi Kornicker, 1983	+F, M	2–12	+
H. tenax Kornicker, 1995	- F	400	-
H. thrix Kornicker, 1992	+ F, IVF	13–20	+
H. vatrax Kornicker, 1995	+ F, M	204	+
H. vix Kornicker, 1991	- F	3–7	+
H. vortex Kornicker, 1995	- F	363	+
H. species A: Kornicker, 1978	- F	73–102	+
H. species B: Kornicker, 1978	- F	140	+

* Kornicker (1992: 108) determined that juvenile males have lateral eyes correcting the description of the species by Kornicker (1978: 39).

** "Sac with many minute yellow cells in vicinity where lateral eye generally occurs may be lateral eye." (Kornicker & Caraion, 1977: 31). This was interpreted not to be an eye in Kornicker (1992: Table 12). It is interpreted here to be a weakly developed eye.